T0135815

Monitoring Procedures

for

Detecting Gradual Changes

INAUGURAL-DISSERTATION

zur

Erlangung des Doktorgrades

der Mathematisch-Naturwissenschaftlichen Fakultät

der Universität zu Köln

vorgelegt von

Hella Christine Timmermann

aus Bergisch-Gladbach

Februar 2014

Berichterstatter/in: Prof. Dr. Josef G. Steinebach, Universität zu Köln
Prof. Dr. Wolfgang Wefelmeyer, Universität zu Köln
Prof. Dr. Marie Hušková, Karls-Universität, Prag

Tag der letzten mündlichen Prüfung: 21. Mai 2014

Bibliografische Information der Deutschen Nationalbibliothek

Die Deutsche Nationalbibliothek verzeichnet diese Publikation in der
Deutschen Nationalbibliografie; detaillierte bibliografische Daten sind
im Internet über http://dnb.d-nb.de abrufbar.

ISBN 978-3-8325-3756-2

Logos Verlag Berlin GmbH
Comeniushof, Gubener Str. 47,
10243 Berlin
Tel.: +49 (0)30 42 85 10 90
Fax: +49 (0)30 42 85 10 92
INTERNET: http://www.logos-verlag.de

Abstract

This thesis is concerned with the sequential detection of gradual changes in the location of a stochastic process in two different settings: In Chapter 1 we consider a general stochastic process with a linear drift term which exhibits a possible gradual (non-linear) perturbation at some unknown time point. In Chapter 2 we approach the question of how to detect a gradual change in the location of an unobservable (renewal) process based on observations of the corresponding counting process. We suggest to base the inference on the inverse of the counting process, which behaves similarly as the underlying process itself. In both settings, we introduce detectors and stopping times which follow a common approach on detecting gradual changes (see e.g. Jarušková (1998) and Hušková (1998a)): Essentially the idea is to introduce a weight function in order to put less weight on early observations - where a possible change has either not occurred (yet) or is still quite small - and heavy weight on late observations - where a possible change is at its current maximum. This idea is further supported by a (quasi) maximum likelihood approach which suggests to use the assumed type of gradual change as a weighting. Via asymptotic results under the null hypothesis we obtain critical values for the suggested procedures. Under the alternative we show the consistency of the procedures as well as the asymptotic normality of the (standardized) delay times, i.e. the time lag between a change point and its detection.

Zusammenfassung

Die vorliegende Arbeit befasst sich mit der sequentiellen Aufdeckung gradueller Änderungen im Erwartungswert eines stochastischen Prozesses in zwei unterschiedlichen Situationen: In Kapitel 1 untersuchen wir einen allgemeinen stochastischen Prozess mit einem linearen Drift, der zu einem unbekannten Zeitpunkt eine graduelle (nichtlineare) Störung aufweist. In Kapitel 2 beschäftigen wir uns mit der Frage, wie man eine graduelle Änderung im Erwartungswert eines unbeobachtbaren (Erneuerungs-) Prozesses basierend auf Beobachtungen des zugehörigen Zählprozesses aufdecken kann. Wir verfolgen den Ansatz, die Untersuchungen mit Hilfe des zum Zählprozess inversen Prozesses durchzuführen, welcher sich ähnlich verhält wie der ursprüngliche Prozess selbst. In beiden Szenarien stellen wir Detektoren und Stoppzeiten vor, die einem allgemein üblichen Ansatz zur Aufdeckung gradueller Änderungen folgen (siehe zum Beispiel Jarušková (1998) und Hušková (1998a)): Die wesentliche Idee besteht darin eine Gewichtsfunktion einzuführen, mittels derer wir wenig Gewicht auf frühe Beobachtungen - in denen eine mögliche Änderung entweder (noch) nicht eingetreten ist oder bisher nur sehr klein ist - und stärkeres Gewicht auf die neuen Beobachtungen - wo eine mögliche Änderung zum aktuellen Zeitpunkt maximal ist - legen. Diese Idee wird zudem durch einen (quasi) Maximum Likelihood Ansatz gestützt, welcher empfiehlt, die erwartete Art der Änderung als Gewichtung zu verwenden. Mittels asymptotischer Resultate unter der Nullhypothese können wir kritische Werte für die vorgeschlagenen Prozeduren herleiten. Unter der Alternative zeigen wir sowohl die Konsistenz der Testverfahren als auch die asymptotische Normalität der (standardisierten) Verzögerungszeiten, sprich der Zeit zwischen dem Auftreten und der Aufdeckung einer Änderung.

Acknowledgment

I am deeply grateful to my supervisor Prof. Steinebach for putting the trust in me and giving me the opportunity to write this thesis and for helping me with many smaller or larger problems throughout this time. Furthermore, I owe special thanks to Prof. Wefelmeyer for the many interesting discussions at lunch and to Prof. Hušková for giving me the opportunity to visit her in Prague and for giving me advice at the beginning of my studies. Both of them have kindly agreed to review my thesis. I shall not forget to thank all of the members of the current and former stochastic group for the many enjoyable lunches and coffee breaks, especially to Stefan Mihalache who made me feel very welcome during my first year as a PhD-student.

My very sincere thank you goes to all my friends who put up with me during good and bad times and who always let me know that not everything is about mathematics. Last but not least I would like to thank Leo - I would not have been able to do this without you!

... für meine Mutter und meinen Vater ...

Contents

Conventions

The following notations will be used frequently:

For any $x \in \mathbb{R}$ let

$$x_+ := \max\{x, 0\},$$
$$\log(x) := \ln\left(\max\{x, \exp(1)\}\right),$$
$$\text{sgn}(x) := \begin{cases} 1 & \text{if } x > 0, \\ -1 & \text{if } x < 0, \\ 0 & \text{if } x = 0, \end{cases}$$

and $[x]$ be the integer part of x.

For any two real-valued sequences $\{x_n\}_{n \in \mathbb{N}}$ and $\{y_n\}_{n \in \mathbb{N}}$ we denote

$$x_n \sim y_n \;:\Leftrightarrow\; x_n/y_n \to 1 \text{ as } n \to \infty,$$
$$x_n \simeq y_n \;:\Leftrightarrow\; \exists\, c > 0 : x_n/y_n \to c \text{ as } n \to \infty.$$

For any two sequences $\{X_n\}_{n \in \mathbb{N}}$ and $\{Y_n\}_{n \in \mathbb{N}}$ of real-valued random variables we denote

$$X_n \overset{\text{a.s.}}{\sim} Y_n \;:\Leftrightarrow\; X_n/Y_n \overset{\text{a.s.}}{\longrightarrow} 1 \text{ as } n \to \infty,$$
$$X_n \overset{\text{a.s.}}{\simeq} Y_n \;:\Leftrightarrow\; \exists\, c > 0 : X_n/Y_n \overset{\text{a.s.}}{\longrightarrow} c \text{ as } n \to \infty.$$

Further, let $\min \emptyset := \infty =: \inf \emptyset$.

By φ and ϕ we denote the density and distribution functions of the standard normal distribution.

Introduction

A lot of decisions made nowadays (e.g. in finance, insurance or medicine) rely on predictions made by statistical methods. Often, the underlying models depend on parameters which are either known or estimated from past experience and assumed to be constant over time. The aim of change point analysis is essentially to survey the stability of model parameters in a given time series and, if necessary, to reveal possible changes in those parameters.

The origins of change point analysis date back to Page (1954) who introduced a control chart for the purpose of quality control. Since then, change point analysis has developed into a wide research field with applications wherever statistical techniques rely on the stability of certain patterns over time.

Probably the most studied model is a location model with a possible (single) sudden jump in the mean: Given n independent random variables $X_1, \ldots, X_n$ with means μ_i, such that $X_i - \mu_i$ are identically distributed, one wishes to test the null hypotheses

$$H_0 : \mu_1 = \ldots = \mu_n$$

against the alternative

$$H_1 : \mu_1 = \ldots = \mu_{k^*-1} \neq \mu_{k^*} = \ldots = \mu_n,$$

where the change point k^* and $\mu_1 \neq \mu_{k^*}$ are unknown.

Many results on this problem rely on almost sure approximations of the partial sums $\left\{ \sum_{i=1}^{k} X_i - \mu_i \mid k \in \mathbb{N} \right\}$ by a Wiener process. An invariance principle of this type was first proposed by Strassen (1964), who constructed a Wiener process which approximates the partial sums by the convergence rate of the law of the iterated logarithm under the assumption of $E|X_1|^2 < \infty$. His work paved the way for other strong approximations, resulting in the seminal KMT-construction (see Komlós et al. (1975), (1976)), which gives us the best possible rate of such a strong approximation under the assumptions of $E|X_i|^{1/\kappa} < \infty$ for some $0 < \kappa < 1/2$. A general overview on this topic can be found e.g. in Csörgő and Révész (1981).

The approximation by a Wiener process motivates a derivation of a test statistic via a maximum likelihood approach, where (for a moment) we assume X_i to be

1

independently $N(\mu_i, \sigma^2)$–distributed for some $\sigma^2 > 0$. Denoting $\varphi(x)$ as the density of the standard normal distribution one rejects the null hypothesis if

$$\ln\left(\frac{\sup_\mu \prod_{i=1}^{k^*} \varphi(X_i - \mu) \ \sup_{\tilde{\mu} \neq \mu} \prod_{j=k^*+1}^{n} \varphi(X_j - \tilde{\mu})}{\sup_\mu \prod_{i=1}^{n} \varphi(X_i - \mu)}\right)$$

is large. Since k^* is unknown the expression is further maximized over all possible change points (i.e. all $k = 1, \ldots, n-1$), which yields a prominent cumulative sum (CUSUM) statistic

$$\max_{k=1,\ldots,n-1} \frac{\sum_{i=1}^{k} \left(X_i - \sum_{j=1}^{n} X_j/n\right)}{\sqrt{(k(n-k))/n}}.$$

Many results have been published on this test statistic and various modifications have been proposed. For a general overview we refer to Csörgő and Horváth (1997) and Aue and Horváth (2013).

In many applications, however, not a sudden but a gradual change, i.e. a slowly increasing or decreasing change, in the location of a process seems to be the far more realistic scenario. For instance (if no obvious shock has occurred) meteorological parameters, such as global temperature, elevation of lakes or average precipitation, are more likely to change slowly than abruptly. Still, testing procedures that are designed to detect gradual changes in the location of a process have gotten less attention in the literature. Jarušková (1998) and Hušková (1998a), (1998b) started to study models with linearly increasing changes at k^*, i.e. (in the latter case)

$$X_i = \varepsilon_i + \delta_n \left(\frac{i - k^*}{n}\right)^\gamma_+, \quad i = 1, \ldots, n,$$

where $\gamma = 1$, $\delta_n \neq 0$ and $\varepsilon_i \sim (\mu, \sigma^2)$ being independently, identically distributed. A generalization of this model, allowing for a slope parameter $\gamma > 0$ was further analyzed by Hušková (1999), Hušková and Steinebach (2000), under weak dependence by Steinebach (2000) and via a permutation approach by Kirch and Steinebach (2006). Arguing as in the case of an abrupt change (via a quasi maximum likelihood approach) one rejects the null hypothesis of "no change" if

$$\max_{k=1,\ldots,n-1} \ln\left(\frac{\sup_\mu \prod_{i=1}^{k} \varphi(X_i - \mu) \ \sup_{\delta_n \neq 0} \prod_{j=k+1}^{n} \varphi(X_j - \mu - \delta_n((j-k^*)/n)^\gamma)}{\sup_\mu \prod_{i=1}^{n} \varphi(X_i - \mu)}\right)$$

is large, which leads to the following weighting of the observations in the test statistic:

$$\max_{k=1,\ldots,n-1} \frac{\sum_{i=1}^{n}((i-k)/n)^\gamma_+ \left(X_i - \sum_{j=1}^{n} X_j/n\right)}{\sqrt{\sum_{i=1}^{n}\left(((i-k)/n)^\gamma_+ - \left(\sum_{j=1}^{n}((j-k)/n)^\gamma_+\right)/n\right)^2}}.$$

Heuristically speaking, one puts heavier weights on later observations - where a possible change will be currently maximal - in order to detect a change faster than by putting the same weight on every observation.

All of the procedures mentioned so far evaluate a given time series retrospectively, i.e. after the n-th observation, having the disadvantage that a perturbed process might have kept on running for long before the disorder is detected. Chu et al. (1996), on the other hand, led the way to a sequential approach in change point analysis: The idea is to decide with each newly made observation whether the null hypothesis of no change is still justified or whether a significant change in the structure (i.e. in their case in the coefficients of a linear regression model) has occurred. The idea inspired further research in this area, among many others Aue (2003), Horváth et al. (2004), Berkes et al. (2004), Kühn (2008) and Mihalache (2012).

Sequential (or monitoring) procedures can be subdivided into two classes: closed-end and open-end procedures. For closed-end procedures one prescribes a fixed truncation point after which the monitoring of the process terminates even if no change has been detected until then. Typically, the test statistic depends on the truncation point, hence the length of the monitoring period has to be known in advance. Open-end procedures, on the contrary, have no fixed sample sizes, which has the benefit that one does not have to decide in advance for how long the process shall be monitored.

Chapter 1 of this thesis resumes the work by Steinebach and Timmermann (2011), where a closed-end procedure for a (generalized) sequential version of the gradual change model from above is introduced. In this thesis, we derive an open-end procedure for a general stochastic process with a possible gradual change in the location parameter and analyze its properties under the null hypothesis and under the alternative. Chapter 2 deals with structural breaks in counting processes. We investigate a scenario where a gradual change appears in the location of the underlying (e.g. partial sum) process, which is typically unobservable. We introduce a test statistic based on the inverse of the counting process, which behaves similarly as the underlying process itself, establish closed-end and open-end monitoring procedures in this setting and study their properties.

Chapter 1

Monitoring general gradual changes

In this chapter we consider a general stochastic process with a linear trend which exhibits a possible perturbation at some unknown time point, where we also allow for a possible change in the scale parameter. We aim to keep the setting as general as possible, focusing on the required invariance principles, yet, throughout this chapter one should always keep in mind the standard example of

$$Z(k) = \sum_{i=1}^{k} \Big(\varepsilon_i + \mu + \delta_m ((i - T^*)/m)_+^{\gamma} \Big), \quad k \in \mathbb{N}$$

where $\mu, \gamma, \delta_m \in \mathbb{R}$, with $\gamma > 0$, $\delta_m \neq 0$, $T^* > m$ and $\{\varepsilon_i \,|\, i \in \mathbb{N}\}$ being independently, identically distributed random variables, which is a sequential version of the model investigated by Hušková (1998a) (see Example 1.1.1 for the precise setting).

This chapter is organized as follows: The testing problem is introduced in detail in Section 1.1 and suitable detectors and stopping times, designed to detect the respective alternatives, are suggested in Section 1.2. Section 1.3 deals with the asymptotic behavior of the stopping times under the null hypothesis. The results can be used to adjust the stopping times such that the procedures attain a prescribed $\alpha-$level asymptotically. In the two sections following upon, we study the behavior of the stopping times under the assumption that a change took place. In Section 1.4 we show the consistency of the suggested procedures and develop a first approach on approximating the delay time, i.e. the time lag between the change point and its detection. Under stronger assumptions (namely in case of an "early" change point) one can also show asymptotic normality of standardized delay times, which is done in Section 1.5. Finally, the finite sample behavior of the monitoring procedures is illustrated in a small simulation study given in Section 1.6 and some minor calculations are postponed to Section 1.7.

Our detectors will make use of the so called in-control parameters, i.e. the parameters which are constant ("in control") under the null hypothesis but not constant ("out of control") under the alternative. Typically, the in-control parameters are unknown and have to be estimated, so each section is again divided into a subsection which contains the results for known in-control parameters and another subsection, where the results are transferred to the case of estimated in-control parameters.

1.1 Setting of the problem

Assume, we sequentially observe a process

$$Z(t) = \begin{cases} bY(t) + at, & 0 \le t \le T^*, \\ bY(T^*) + b^* Y^*(t - T^*) + at + \Delta_{m,\gamma}(t - T^*), & T^* < t < \infty, \end{cases} \quad (1.1.1)$$

at integer time points (i.e. we monitor $Z(0), Z(1), Z(2), \ldots$) where $\{Y(t) \mid t \ge 0\}$ and $\{Y^*(t) \mid t \ge 0\}$ are two stochastic processes for which the following invariance principles shall hold true: There are Wiener processes $\{W(t) \mid t \ge 0\}$ and $\{W^*(t) \mid t \ge 0\}$ and some $0 < \kappa < 1/2$ such that

$$\sup_{0 < t < T^*} \frac{|Y(t) - W(t)|}{t^\kappa} = \mathcal{O}_P(1), \quad (1.1.2)$$

$$\sup_{0 < t < \infty} \frac{|Y^*(t) - W^*(t)|}{(t + T^*)^\kappa} = \mathcal{O}_P(1). \quad (1.1.3)$$

For the sake of generality, we only assume a weak approximation of $Y(t)$ and $Y^*(t)$. However, in many cases, where such weak approximations are known, they were in fact deduced from strong approximations.

Further, let $a, b, b^* \in \mathbb{R}$ where $b, b^* > 0$. The "change" $\Delta_{m,\gamma}(t - T^*)$ shall be a strictly in- or decreasing, deterministic function in $t \ge T^*$, possibly relying on further parameters (indicated by γ) and on the length m of a so called "training period", i.e. an observation period during which we know that no change occurs. The assumption of such a training period is known as "non-contamination assumption" and formally states that $m < T^*$, for all $m \in \mathbb{N}$. All of the asymptotic results presented in this thesis rely on $m \to \infty$. Moreover, for the sake of simplicity, we assume $Y(0) = Y^*(0) = 0$ a.s. and $\Delta_{m,\gamma}(t - T^*) = 0$ for $t \le T^*$.

We are interested in testing the null hypothesis

$H_0\colon\ T^* = \infty$ "no change"

against either one of the following alternatives

$H_1^+\colon\ T^* < \infty,\ \Delta > 0$ "one-sided, positive change",
$H_1^-\colon\ T^* < \infty,\ \Delta < 0$ "one-sided, negative change",
$H_1\ :\ T^* < \infty,\ \Delta \ne 0$ "two-sided change",

where by $\Delta > 0\ (< 0, \ne 0)$ we mean $\Delta_{m,\gamma}(t - T^*) > 0\ (< 0, \ne 0)$ for all $t > T^*$.

In the following examples we describe two processes $\{Z(t) \mid t \ge 0\}$ which fit into the above framework. Further examples can be found e.g. following Horváth and Steinebach (2000) and Steinebach (2000).

Example 1.1.1. *Assume we observe a process*

$$X_i = \varepsilon_i + \mu + \delta_m \left(\frac{i - T^*}{m} \right)_+^\gamma,$$

where $\mu, \gamma, \delta_m \in \mathbb{R}$, with $\gamma > 0$, $\delta_m \neq 0$ and $\{\varepsilon_i \,|\, i \in \mathbb{N}\}$ being independently, identically distributed random variables with $E(\varepsilon_i) = 0$, $\mathrm{Var}(\varepsilon_i) = \sigma^2 > 0$ and $E|\varepsilon_i|^{1/\kappa} < \infty$ for some $0 < \kappa < 1/2$. The parameter $\delta = \delta_m$ may particularly depend on m, where typically one is interested in so called "local alternatives", i.e. $\delta_m \to 0$ as $m \to \infty$. On setting

$$a = \mu, \; b = b^* = \sigma,$$

$$Y(t) = \sum_{i=1}^{[t]} \varepsilon_i,$$

$$Y^*(t) = \sum_{i=[T^*]+1}^{[t]} \varepsilon_i,$$

$$\Delta_{m,\gamma}(t - T^*) = \sum_{i=1}^{[t]} \delta_m \left(\frac{i - T^*}{m} \right)_+^\gamma$$

we obtain a process in the form of (1.1.1). Further, by Komlós et al. (1975) we know that there is some Wiener process $\{W(t) \,|\, t \geq 0\}$ such that $|\sum_{i=1}^n \varepsilon_i - \sigma W(n)| \stackrel{a.s.}{=} \mathcal{O}(n^\kappa)$, hence on approximating $Y(t)$ by $W(t)$ and $Y^*(t)$ by $W^*(t) := W(t + T^*) - W(T^*)$ the invariance principles (1.1.2) and (1.1.3) are fulfilled. (In fact, (1.1.2) and (1.1.3) are even fulfilled if we replace $\mathcal{O}_P(1)$ by $\mathcal{O}(1)$ a.s.)

Example 1.1.2. *Again, assume we observe*

$$X_i = \varepsilon_i + \mu + \delta_m \left(\frac{i - T^*}{m} \right)_+^\gamma$$

with μ, γ, δ_m as above but ε_i being a linear process, i.e.

$$\varepsilon_i = \sum_{k=0}^\infty a_k \, e_{i-k},$$

where for e_i it holds that

- · *$\{e_i \,|\, i \in \mathbb{Z}\}$ are independently, identically distributed random variables with $E(e_i) = 0$, $\mathrm{Var}(e_i) = \sigma^2 > 0$ and $E|e_i|^{1/\tilde{\kappa}} < \infty$ for some $0 < \tilde{\kappa} < 1/2$,*

- · *the random variables are smooth with density function f satisfying*

$$\sup_{-\infty < s < \infty} \frac{1}{|s|} \int_{-\infty}^\infty |f(t + s) - f(t)| \, dt < \infty$$

and for the coefficients a_k it holds that

· $a_k = \mathcal{O}\big(k^\beta\big)$ for some $\beta < 3/2$, as $k \to \infty$,

· $\sum_{k=0}^{\infty} a_k\, z^k \neq 0$ for all $z \in \mathbb{C}$ with $|z| < 1$,

· $\eta^2 := \sigma^2 \bigg(\sum_{k=0}^{\infty} a_k \bigg)^2 > 0.$

Lemma 2.2 of Horváth (1997) states that there is some Wiener process $\{W(t)\,|\,t \geq 0\}$ such that $\big| \sum_{i=1}^{n} \varepsilon_i - \eta\, W(n) \big| \overset{a.s.}{=} o\big(n^\kappa\big)$ for some $0 < \kappa < 1/2$, as $n \to \infty$. On setting a, $Y(t)$, $Y^*(t)$, $\Delta_{m,\gamma}(t)$, $W(t)$ and $W^*(t)$ as in Example 1.1.1 and $b = b^* = \eta$ we obtain a process as in (1.1.1), which fulfills the invariance principles (1.1.2) and (1.1.3). (In fact, (1.1.2) and (1.1.3) are even fulfilled if we replace $\mathcal{O}_P(1)$ by $\mathcal{O}(1)$ a.s.)

1.2 Stopping times

For our sequential testing procedure we are looking for stopping times τ_m such that under the null hypothesis it holds for some fixed $\alpha \in (0,1)$ that

$$\lim_{m \to \infty} P(\tau_m < \infty) = \alpha$$

and under the alternative it holds that

$$\lim_{m \to \infty} P(\tau_m < \infty) = 1.$$

The stopping times suggested below are based on detectors that are evaluated with each newly made observation to determine whether a suitably chosen threshold function is exceeded (in which case the sequential procedure is stopped) or not (in which case the monitoring of the process is continued).

We standardize the observations by the in-control parameters, i.e. (in our case) the location-parameter a and the scale-parameter b. If the in-control parameters are unknown, which is usually the case, they have to be estimated. Detectors and stopping times for known or unknown in-control parameters are introduced in Section 1.2.1 or Section 1.2.2, respectively.

1.2.1 Stopping times for known in-control parameters

We pick up the approach of Jarušková (1998) and Hušková (1998a) constructing our detectors as sums of weighted, standardized increments of the observed time series, putting the heaviest weight on the latest observation, where the "size" of a possible change is the largest. Denoting the increments by $Z_i = Z(i) - Z(i-1)$ we consider the following detectors:

$$T_k = \frac{\sum_{i=1}^{k} g(i/m) \, (Z_i - a)}{b \, \sqrt{m}}, \quad k \geq m, \tag{1.2.1}$$

where m is a training period after which the sequential testing procedure shall start. We present our results for a general weight function g which is further specified in Assumption 1.2.1 below. However, throughout this chapter the standard example for a weight function will be $g(t) = t^\lambda$ for some $\lambda > 0$, where for $\Delta_{m,\gamma}$ as in Example 1.1.1 and Example 1.1.2 a (quasi) maximum likelihood approach suggests $\lambda = \gamma$ as a proper choice (see Remark 1.2.2).

If, for a moment, we assume the increments Z_i of the considered process to be independently, identically distributed with mean a and variance b^2, we have

$$\mathrm{Var}\left(T_k\right) = \frac{\sum_{i=1}^{k} g^2(i/m) \, \mathrm{Var}(Z_i)}{b^2 \, m} = \frac{\sum_{i=1}^{k} g^2(i/m)}{m} \approx \int_0^{k/m} g^2(x) \, dx.$$

Hence, we introduce the function

$$G(t) = \int_0^t g^2(x) \, dx, \tag{1.2.2}$$

which we will use to standardize the detectors. In fact, $G(t)$ is the variance of the process $\left\{ \int_0^t g(x)\, dW(x) \,|\, t \geq 1 \right\}$, which will turn out to be the limit of $\{T_k \,|\, k \geq m\}$ (see Theorem 1.3.1 below for details).

The idea is to stop the testing procedure as soon as the detectors exceed some critical threshold function h, so our stopping times are

$$\tau_m^+ = \min\left\{ k = m, \dots, \infty \;\middle|\; T_k/h(G(k/m)) > 1 \right\},$$
$$\tau_m^- = \min\left\{ k = m, \dots, \infty \;\middle|\; -T_k/h(G(k/m)) > 1 \right\},$$
$$\tau_m = \min\left\{ k = m, \dots, \infty \;\middle|\; |T_k|/h(G(k/m)) > 1 \right\},$$

where τ_m^+ shall detect the one-sided, positive alternative H_1^+, τ_m^- shall detect the one-sided, negative alternative H_1^- and τ_m shall detect the two-sided alternative H_1. In all cases the threshold function h has to be chosen in such a way that the test attains a prescribed level α asymptotically. Possible threshold functions are given in Remark 1.2.2 below.

For our asymptotic results we need the following set of assumptions on the weight and threshold functions:

Assumption 1.2.1. *Let $g : [0, \infty) \to [0, \infty)$ be an increasing and continuous function which is differentiable on $(0, \infty)$ and (for the sake of simplicity) satisfies $g(0) = 0$. Let $h = h_c : [G(1), \infty) \to (0, \infty)$ be an additional increasing (threshold) function such that the following growth conditions hold true, as $m \to \infty$,*

$$\sup_{t \geq m} \frac{t^\kappa \, g(t/m)}{\sqrt{m}\, h(G(t/m))} \to 0, \qquad\qquad (1.2.3)$$

$$\sup_{t \geq m} \sup_{0 \leq \xi \leq 1/m} \frac{g'(t/m - \xi)\sqrt{t \, \log\log(t/m)}}{m^{3/2}\, h(G(t/m))} \to 0, \qquad\qquad (1.2.4)$$

$$\sup_{t \geq m} \left| 1 - \frac{h(G(t/m))}{h(G([t]/m))} \right| \to 0, \qquad\qquad (1.2.5)$$

where κ is defined in (1.1.2) and (1.1.3) and $G(t)$ is defined in (1.2.2). Further, the threshold function h shall be chosen such that $\sup_{t \geq G(1)} |W(t)|/h(t)$ has a non-degenerated distribution.

Remark 1.2.2. *Possible choices of weight and threshold functions are $g(t) = t^\lambda$, for some $\lambda > 0$, which implies $G(t) = t^{1+2\lambda}/(1+2\lambda)$, and $h = h_c^{(i)}$ where*

$$h_c^{(1)}(t) = c\,t, \quad c > 0,$$
$$h_c^{(2)}(t) = \sqrt{t}\, f^{-1}(\ln(t) + f(c)), \quad \text{where } f(t) = t^2 + 2\ln(\phi(t)), \; c > 0,$$
$$h_c^{(3)}(t) = \sqrt{t\,(c^2 + \ln(t))}, \quad \ln(G(1)) > -c^2.$$

The functions $h_c^{(1)}(t)$ and $h_c^{(2)}(t)$ are suggested for the one-sided stopping times τ_m^+ and τ_m^- and $h_c^{(3)}(t)$ is applicable for the two-sided stopping time τ_m (see Remark 1.2.3). The growth conditions of Assumption 1.2.1 are verified in Section 1.7 and the fact that $\sup_{t \geq G(1)} |W(t)|/h(t)$ has a non-degenerated distribution follows from Remark 1.2.3.

On choosing one of the threshold functions suggested in Remark 1.2.2 one can make use of the following distributions, shown by Robbins and Siegmund (1970) (see Examples 1, 2 and 3 therein), to adjust the testing procedures such that they attain a prescribed level α asymptotically (cf. Theorem 1.3.1 and Theorem 1.3.4 below).

Remark 1.2.3. *Let* $\{W(t)\,|\,t \geq 0\}$ *be a Wiener process and* $f(t) = t^2 + 2\ln(\phi(t))$.

1. *It holds for all* $c > 0$ *and* $t_0 > 0$ *that*

$$P\left(\max_{t \geq t_0} \frac{W(t)}{c\,t} < 1\right) = 2\phi(c\sqrt{t_0}) - 1.$$

2. *It holds for* $f(t) = t^2 + 2\ln(\phi(t))$, $c > 0$ *and* $t_0 = 1$ *that*

$$P\left(\sup_{t \geq t_0} \frac{W(t)}{\sqrt{t}\; f^{-1}(\ln(t) + f(c))} < 1\right) = \phi(c) - \varphi(c)\,(c + \varphi(c)/\phi(c)).$$

A general expression of the distribution for arbitrary $t_0 > 0$ *is given in Example 2 of Robbins and Siegmund (1970).*

3. *It holds for all* $c > 0$ *and* $\ln(t_0) > -c^2$ *that*

$$P\left(\sup_{t \geq t_0} \frac{|W(t)|}{\sqrt{t(c^2 + \ln(t))}} < 1\right)$$
$$= 2\phi\big(\sqrt{c^2 + \ln(t_0)}\big) - 2\varphi(c)\sqrt{(c^2 + \ln(t_0))/t_0} - 1.$$

1.2.2   Stopping times for unknown in-control parameters

The in-control parameters a and b are usually unknown, hence (in this case) they have to be estimated. On estimating a we consider the empirical mean

$$\hat{a}_k = \frac{1}{k} \sum_{i=1}^{k} \big(Z(i) - Z(i-1) \big) = \frac{Z(k)}{k}, \quad k \geq m, \tag{1.2.6}$$

whereas on estimating b we consider a general (possibly sequential) estimate $\hat{b}_k$ which satisfies under the null hypothesis

$$\sup_{k=m,\dots,\infty} |\hat{b}_k - b| = o_P(1), \quad k \geq m. \tag{1.2.7}$$

A possible choice for $\hat{b}_k$ is given in Remark 1.2.6 below. Replacing the unknown in-control parameters by their estimates we obtain the following detectors:

$$\hat{T}_k = \frac{\sum_{i=1}^{k} g(i/m)\,(Z_i - \hat{a}_k)}{\hat{b}_k \,\sqrt{m}}, \quad k \geq m.$$

If, for a moment, we assume the increments $Z_i = Z(i) - Z(i-1)$ of the considered process to be independently, identically distributed with mean a and variance b^2, we have

$$\mathrm{Var}\big(\hat{T}_k\big) \approx \frac{\sum_{i=1}^{k} g^2(i/m)}{m} - \frac{\big(\sum_{i=1}^{k} g(i/m) \big)^2}{m\,k}$$

$$\approx \int_{0}^{k/m} g^2(x)\,dx - \frac{\big(\int_{0}^{k/m} g(x)\,dx \big)^2}{k/m}.$$

Hence, analogously to (1.2.2), we introduce the function

$$\tilde{G}(t) = \int_{0}^{t} g^2(x)\,dx - \frac{\big(\int_{0}^{t} g(x)\,dx \big)^2}{t}, \tag{1.2.8}$$

with $\tilde{G}(0) := \lim_{t \to 0} \tilde{G}(t) = 0$, which we will use to standardize our detectors. In fact, $\tilde{G}(t)$ is the variance of the process $\{U(t)\,|\,t \geq 1\}$, (where $U(t)$ is defined in (1.3.14) below), which will turn out to be the limit of $\{\hat{T}_k\,|\,k \geq m\}$ (see Theorem 1.3.4 below for details).

We consider the following modified stopping times:

$$\hat{\tau}_m^+ = \min \big\{ k = m, \dots, \infty \;\big|\; \hat{T}_k / h(\tilde{G}(k/m)) > 1 \big\},$$
$$\hat{\tau}_m^- = \min \big\{ k = m, \dots, \infty \;\big|\; -\hat{T}_k / h(\tilde{G}(k/m)) > 1 \big\},$$
$$\hat{\tau}_m = \min \big\{ k = m, \dots, \infty \;\big|\; |\hat{T}_k| / h(\tilde{G}(k/m)) > 1 \big\},$$

where, as before, τ_m^+ shall detect the one-sided, positive alternative H_1^+, τ_m^- shall detect the one-sided, negative alternative H_1^- and τ_m shall detect the two-sided alternative H_1. In all cases, the threshold function h is to be chosen in such a way, that the test attains a prescribed level α asymptotically. Possible threshold functions are given in Remark 1.2.5 below.

For our asymptotic results we need the following set of assumptions on the weight and threshold functions:

Assumption 1.2.4. Let $g : [0, \infty) \rightarrow [0, \infty)$ be an increasing and continuous function which is differentiable on $(0, \infty)$ and (for the sake of simplicity) satisfies $g(0) = 0$ and $g'(t) \geq \tilde{c} \, t^{-3/2}$ for some $\tilde{c} > 0$ and for all $t \geq 1$. Let $h = h_c : [\tilde{G}(1), \infty) \rightarrow (0, \infty)$ be an additional increasing (threshold) function such that the following growth conditions hold true, as $m \rightarrow \infty$,

$$\sup_{t \geq m} \frac{t^\kappa \, g(t/m)}{\sqrt{m} \, h(\tilde{G}(t/m))} \rightarrow 0, \tag{1.2.9}$$

$$\sup_{t \geq m} \sup_{0 \leq \xi \leq 1/m} \frac{g'(t/m - \xi)\sqrt{t \log \log(t/m)}}{m^{3/2} \, h(\tilde{G}(t/m))} \rightarrow 0, \tag{1.2.10}$$

$$\sup_{t \geq m} \left| 1 - \frac{h(\tilde{G}(t/m))}{h(\tilde{G}([t]/m))} \right| \rightarrow 0, \tag{1.2.11}$$

where κ is defined in (1.1.2) and (1.1.3) and $\tilde{G}(t)$ is defined in (1.2.8). Further, the threshold function h shall be chosen such that $\sup_{t \geq \tilde{G}(1)} |W(t)|/h(t)$ has a non-degenerated distribution.

Remark 1.2.5. Possible choices of weight and threshold functions are $g(t) = t^\lambda$, for some $\lambda > 0$, which implies $\tilde{G}(t) = \tilde{\lambda} \, t^{1+2\lambda}$, where $\tilde{\lambda} = \lambda^2 / ((1 + 2\lambda)(1 + \lambda)^2)$, and h as in Remark 1.2.2. The growth conditions of Assumption 1.2.4 are verified in Section 1.7 and the fact that $\sup_{t \geq \tilde{G}(1)} |W(t)|/h(t)$ has a non-degenerated distribution follows from Remark 1.2.3.

Remark 1.2.6. A possible choice for an estimate that fulfills (1.2.7) under the null hypothesis is

$$\hat{b}_k^2 = \frac{1}{\tilde{k}(k - \tilde{k} + 1)} \sum_{j=\tilde{k}}^{k} \left(Z(j) - Z(j - \tilde{k}) - \hat{k} \, Z(k)/k \right)^2, \quad k \geq m,$$

where $\hat{k} = k^q$ for some $2\kappa < q < 1$ (see Section 1.3.3 below). In order to preserve the consistency under the alternative one can also estimate b non-sequentially, i.e. not with every newly made observation, but rather by only using the data obtained in the training period.

1.2.3 Variance of the detectors

As illustrated in the previous sections the choice of the weight function g determines the variance of the detectors, i.e. $\mathrm{Var}(T_k) \approx G(k/m)$, where

$$G(t) = \int_0^t g^2(x)\,dx,$$

and $\mathrm{Var}(\hat{T}_k) \approx \tilde{G}(k/m)$, where

$$\tilde{G}(t) = \int_0^t g^2(x)\,dx - \frac{\left(\int_0^t g(x)\,dx\right)^2}{t}$$

with $\tilde{G}(0) = 0$. For our asymptotic results in this chapter we need that $G([1,\infty)) = [G(1),\infty)$, or $\tilde{G}([1,\infty)) = [\tilde{G}(1),\infty)$, respectively. This is verified in the following lemma:

Lemma 1.2.7. *Let* $g: [0,\infty) \to [0,\infty)$ *be an increasing and continuous function, which is differentiable on* $(0,\infty)$ *and satisfies* $g(0) = 0$, *and let* $G(t)$ *and* $\tilde{G}(t)$ *be as above. Then* $G(t)$ *and* $\tilde{G}(t)$ *are increasing, differentiable and positive for all* $t > 0$. *Furthermore, it holds that* $\lim_{t\to\infty} G(t) = \infty$ *and if* $g'(t) \geq \tilde{c}\,t^{-3/2}$ *for some* $\tilde{c} > 0$ *and for all* $t \geq 1$, *it holds that* $\lim_{t\to\infty} \tilde{G}(t) = \infty$.

Proof of Lemma 1.2.7. Obviously, by the fundamental theorem of calculus $G(t)$ and $\tilde{G}(t)$ are differentiable with $G'(t) = g^2(t) > 0$ and

$$
\begin{aligned}
\tilde{G}'(t) &= g^2(t) - \frac{2g(t)t\int_0^t g(x)\,dx - \left(\int_0^t g(x)\,dx\right)^2}{t^2} \\
&= \frac{g^2(t)t^2 - 2g(t)t\int_0^t g(x)\,dx + \left(\int_0^t g(x)\,dx\right)^2}{t^2} \\
&= \left(\frac{g(t)t - \int_0^t g(x)\,dx}{t}\right)^2 > 0
\end{aligned}
$$

for $t > 0$, where the strict inequalities follow by the fact that g is increasing. Hence $G(t)$ and $\tilde{G}(t)$ are increasing and by $G(0) = 0 = \tilde{G}(0)$ we have $G(t) > 0$ and $\tilde{G}(t) > 0$ for all $t > 0$.

To see $G(t) \to \infty$ we note that (again by the fact that $g(t)$ is increasing)

$$G(t) \geq \int_1^t g^2(x)\,dx \geq g^2(1)(t - 1) \to \infty.$$

To see $\tilde{G}(t) \to \infty$ we rewrite

$$\tilde{G}(t) = \int_0^t \left(\frac{g(x)x - \int_0^x g(y)\,dy}{x}\right)^2 dx$$

and give the following lower bound for the square root of the integrand: By the assumption on g' it holds for all $x \geq 2$ that

$$\frac{g(x)x - \int_0^x g(y)\,dy}{x}$$

$$= \frac{g(x)\,x - \int_0^{x/2} g(y)\,dy - \int_{x/2}^x g(y)\,dy}{x}$$

$$\geq \frac{g(x)\,x - g(x/2)\,x/2 - g(x)\,x/2}{x}$$

$$= \frac{g(x) - g(x/2)}{x/2}\,\frac{x}{4}$$

$$= g'(\xi)\,x/4 \quad \text{for some } \xi \in [x/2, x]$$

$$\geq g'(\xi)\,\xi/4$$

$$\geq (\tilde{c}/4)\,\xi^{-1/2}$$

$$\geq (\tilde{c}/4)\,x^{-1/2}.$$

Hence, we have

$$\tilde{G}(t) \geq \frac{\tilde{c}^2}{16} \int_2^t x^{-1}\,dx \rightarrow \infty$$

as $t \rightarrow \infty$. $\qquad\qquad\qquad\qquad\qquad\qquad\qquad\qquad\qquad\qquad\qquad\qquad\quad \square$

Remark 1.2.8. *Note that the condition* $g'(t) \geq \tilde{c}/t^{-3/2}$ *for some* $\tilde{c} > 0$ *and for all* $t \geq 1$ *(see also Assumption 1.2.4) is given for the sake of simplicity. The proof of* $\tilde{G}(t) \rightarrow \infty$ *in fact holds true, if there is some decreasing function* $r :$ $[0, \infty) \rightarrow (0, \infty)$ *such that* $g'(t) \geq \tilde{c}\,r(t)/t$, *for some* $\tilde{c} > 0$ *and* t *sufficiently large, and* $\int_0^\infty r^2(t)\,dt = \infty$. *In Assumption 1.2.4 and Lemma 1.2.7 we considered* $r(t) = t^{-1/2}$.

1.3   Asymptotics under the null hypothesis

In this section we study the asymptotic properties of our test statistics under the null hypothesis for known in-control parameters (Section 1.3.1) and unknown in-control parameters (Section 1.3.2). The results of Theorem 1.3.1 or Theorem 1.3.4, respectively, can be used to adjust the threshold functions such that the monitoring procedures attain a prescribed α-level asymptotically. In Section 1.3.3 we suggest a suitable estimate $\hat{b}_k$ for estimating the scale parameter of the observed process.

1.3.1   Asymptotics for known in-control parameters

Theorem 1.3.1. *Let* $\{W(t) \mid t \geq 0\}$ *be a Wiener process. With the notation and assumptions of Section 1.1 and Section 1.2 it holds under the null hypothesis that*

$$\lim_{m \to \infty} P\big(\tau_m^+ = \infty\big) = P\bigg(\sup_{t \geq G(1)} W(t)/h(t) \leq 1 \bigg),$$

$$\lim_{m \to \infty} P\big(\tau_m^- = \infty\big) = P\bigg(\sup_{t \geq G(1)} W(t)/h(t) \leq 1 \bigg),$$

$$\lim_{m \to \infty} P\big(\tau_m = \infty\big) = P\bigg(\sup_{t \geq G(1)} |W(t)|/h(t) \leq 1 \bigg).$$

A key tool throughout this thesis is the following a.s. bound for the increments of a Wiener process, which was obtained by Csörgő and Révész (1979) and, for the convenience of the reader, is partly restated here. A comprehensive overview on the behavior of Wiener processes, as well as the proof of Proposition 1.3.2 can be found in Csörgő and Révész (1981) (for the latter, see Theorem 1.2.1 therein).

Proposition 1.3.2. *Let* a_T *(* $T \geq 0$ *) be a monotonically non-decreasing function of* T *such that* $0 < a_T \leq T$ *holds true and* T/a_T *is monotonically increasing. Then we have*

$$\limsup_{T \to \infty} \; \sup_{0 \leq t \leq T - a_T} \; \sup_{0 \leq s \leq a_T} \; \beta_T \, |W(t+s) - W(t)| \stackrel{a.s.}{=} 1,$$

where $\beta_T = \big(2 \, a_T \big(\log(T/a_T) + \log\log(T) \big) \big)^{-1/2}$.

Further, we will frequently make use of the following index shift which is known as summation by parts or Abel transformation.

Remark 1.3.3. *For sequences* $\{a_i\}_{i \in \mathbb{N}_0}$ *and* $\{b_i\}_{i \in \mathbb{N}_0}$ *and* $k_1, k_2 \in \mathbb{N}$ *it holds that*

$$\sum_{i=k_1}^{k_2} a_i \, (b_i - b_{i-1}) = \sum_{i=k_1}^{k_2} a_i b_i - \sum_{i=k_1-1}^{k_2-1} a_{i+1} b_i$$

$$= a_{k_2} b_{k_2} - a_{k_1} b_{k_1-1} - \sum_{i=k_1}^{k_2-1} (a_{i+1} - a_i) b_i,$$

which further implies the following two upper bounds: If $\{a_i\}_{i\in\mathbb{N}_0}$ *is increasing, it holds that*

$$\left|\sum_{i=k_1}^{k_2} a_i\,(b_i - b_{i-1})\right| \le |a_{k_2}|\,|b_{k_2} - b_{k_1-1}| + 2\,|a_{k_2} - a_{k_1}| \max_{k=k_1-1,\ldots,k_2} |b_k|,$$

$$\left|\sum_{i=k_1}^{k_2} a_i\,(b_i - b_{i-1})\right| \le 4|a_{k_2}| \max_{k=k_1-1,\ldots,k_2} |b_k|.$$

In the following we give the proof of the main theorem of this section.

Proof of Theorem 1.3.1. The idea of the proof is to make use of the invariance principle (1.1.2) to replace the standardized observations $(Z(k) - ak)/b$ by $W_m(k) := W(k\,m)/\sqrt{m}$ which is by the Brownian scaling property again a Wiener process. The corresponding Gaussian version of the detectors is then replaced by a stochastic integral and the resulting (Gaussian) process can be analyzed via its covariance function.

By (1.1.2) and summation by parts (see Remark 1.3.3) we have

$$
\begin{aligned}
&\sup_{k=m,\ldots,\infty} \left| \frac{T_k}{h(G(k/m))} - \frac{\sum_{i=1}^{k} g(i/m)(W_m(i/m) - W_m((i-1)/m))}{h(G(k/m))} \right| \\
&\le \sup_{k=m,\ldots,\infty} \frac{4\,g(k/m)}{\sqrt{m}\,b\,h(G(k/m))} \sup_{i=0,\ldots,k} |Y(i) - W(i)| \\
&\le \sup_{k=m,\ldots,\infty} \frac{k^{\kappa}\,g(k/m)}{\sqrt{m}\,h(G(k/m))} \mathcal{O}_P(1) \\
&= o_P(1),
\end{aligned}
\tag{1.3.1}
$$

where the last equality follows from (1.2.3). Making use of Remark 1.3.3 once more allows us to replace the sum of the increments of $W_m(t)$ by a stochastic integral:

$$
\begin{aligned}
&\sum_{i=1}^{k} g(i/m)(W_m(i/m) - W_m((i-1)/m)) \\
&= g(k/m)\,W_m(k/m) - \sum_{i=0}^{k-1}(g((i+1)/m) - g(i/m))\,W_m(i/m) \\
&= g(k/m)\,W_m(k/m) - \sum_{i=0}^{k-1} \int_{i/m}^{(i+1)/m} g'(y)\,W_m(i/m)\,dy \\
&= g(k/m)\,W_m(k/m) - \sum_{i=0}^{k-1} \int_{i/m}^{(i+1)/m} g'(y)\,W_m(y)\,dy \\
&\quad + \sum_{i=0}^{k-1} \int_{i/m}^{(i+1)/m} g'(y)\,(W_m(y) - W_m(i/m))\,dy \\
&= \int_{0}^{k/m} g(y)\,dW_m(y) + \sum_{i=0}^{k-1} \int_{i/m}^{(i+1)/m} g'(y)\,(W_m(y) - W_m(i/m))\,dy,
\end{aligned}
\tag{1.3.2}
$$

where in the last line we applied the integration by parts formula for stochastic integrals (see e.g. (3.8) on p.155 of Karatzas and Shreve (1991)). Making use of the a.s. upper bound for the increments of Wiener processes (see Proposition 1.3.2) we see that the second term of (1.3.2) (divided by the threshold function) does not contribute to the asymptotic:

$$
\sup_{k=m,\ldots,\infty} \frac{\left| \sum_{i=0}^{k-1} \int_{i/m}^{(i+1)/m} g'(y)(W_m(y) - W_m(i/m))\, dy \right|}{h(G(k/m))}
$$

$$
\leq \sup_{k=m,\ldots,\infty} \sup_{i=0,\ldots,k-1} \sup_{i/m \leq y \leq (i+1)/m} \frac{|W_m(y) - W_m(i/m)|\; g(k/m)}{h(G(k/m))}
$$

$$
\overset{a.s.}{=} \sup_{k=m,\ldots,\infty} \frac{\sqrt{\log(k)}\; g(k/m)}{\sqrt{m}\; h(G(k/m))}\; \mathcal{O}(1)
$$

$$
= o(1),
$$

where the last equality follows by (1.2.3). On replacing the discrete process by a time continuous one we need to confirm that

$$
\sup_{t \geq m} \left| \frac{\int_0^{t/m} g(y)\, dW_m(y)}{h(G(t/m))} - \frac{\int_0^{[t]/m} g(y)\, dW_m(y)}{h(G([t]/m))} \right| = o_P(1). \tag{1.3.3}
$$

This is shown via the following straightforward decomposition of the above expression:

$$
\sup_{t \geq m} \left| \frac{\int_0^{t/m} g(y)\, dW_m(y)}{h(G(t/m))} - \frac{\int_0^{[t]/m} g(y)\, dW_m(y)}{h(G([t]/m))} \right|
$$

$$
\leq \sup_{t \geq m} \left| \frac{\int_0^{[t]/m} g(y)\, dW_m(y)}{h(G(t/m))} \right| \sup_{t \geq m} \left| 1 - \frac{h(G(t/m))}{h(G([t]/m))} \right|
$$

$$
+ \sup_{t \geq m} \left| \frac{\int_{[t]/m}^{t/m} g(y)\, dW_m(y)}{h(G(t/m))} \right|.
$$

As we proceed in the proof, we will see that the first term on the right hand side is of order $\mathcal{O}_P(1)$ whereas, by (1.2.5), the second term is of order $o(1)$, hence it remains to show that the third term is of order $o_P(1)$. Applying the law of the iterated logarithm and Proposition 1.3.2 to the following

$$
\int_{[t]/m}^{t/m} g(y)\, dW_m(y)
$$

$$
= g(t/m)\, W_m(t/m) - g([t]/m)\, W_m([t]/m) - \int_{[t]/m}^{t/m} g'(y) W_m(y)\, dy \tag{1.3.4}
$$

yields

$$
\begin{aligned}
\sup_{t \geq m} & \frac{\left| \int_{[t]/m}^{t/m} g(y)\, dW_m(y) \right|}{h(G(t/m))} \\
& \leq \sup_{t \geq m} \frac{\left| (g(t/m) - g([t]/m))\, W_m(t/m) \right|}{h(G(t/m))} \\
& \quad + \sup_{t \geq m} \frac{\left| g([t]/m)\, (W_m(t/m) - W_m([t]/m)) \right|}{h(G(t/m))} \\
& \quad + \sup_{t \geq m} \frac{\sup_{0 \leq \xi \leq 1/m} |W_m(t/m - \xi)| \int_{[t]/m}^{t/m} g'(y)\, dy}{h(G(t/m))} \\
& = \sup_{t \geq m} \sup_{0 \leq \xi \leq 1/m} \frac{g'(t/m - \xi)/m \sqrt{t/m \log\log(t/m)}}{h(G(t/m))} \, \mathcal{O}_P(1) \\
& \quad + \sup_{t \geq m} \frac{g([t]/m) \sqrt{\log(t)/m}}{h(G(t/m))} \, \mathcal{O}_P(1) \\
& = o_P(1),
\end{aligned}
$$

where the last equation follows from (1.2.3) and (1.2.4). The two relations above imply (1.3.3), which as a result gives us

$$
\sup_{k=m,\dots,\infty} \frac{T_k}{h(G(k/m))} = \sup_{t \geq m} \frac{\int_0^{t/m} g(y)\, dW_m(y)}{h(G(t/m))} + o_P(1). \tag{1.3.5}
$$

Note that for $0 \leq s \leq t$ it holds true that

$$
\begin{aligned}
\mathrm{Cov} & \left(\int_0^s g(y)\, dW_m(y), \int_0^t g(y)\, dW_m(y) \right) \\
& = \int_0^s g^2(y)\, dy \\
& = \min\{G(s), G(t)\} \\
& = \mathrm{Cov}\left(W(G(s)), W(G(t)) \right),
\end{aligned}
\tag{1.3.6}
$$

hence by the fact that $\left\{ \int_0^t g(y)\, dW_m(y) \,|\, t \geq 0 \right\}$ is a Gaussian process, we have

$$
\left\{ \int_0^t g(y)\, dW_m(y) \,|\, t \geq 0 \right\} \overset{D}{=} \{ W(G(t)) \,|\, t \geq 0 \}.
$$

Finally, note that by Lemma 1.2.7 we have

$$
\sup_{t \geq 1} \frac{W(G(t))}{h(G(t))} = \sup_{t \geq G(t)} \frac{W(t)}{h(t)}.
$$

Combining this with (1.3.5) yields

$$
\sup_{k=m,\dots,\infty} \frac{T_k}{h(G([t]/m))} \overset{D}{\longrightarrow} \sup_{t \geq G(t)} \frac{W(t)}{h(t)},
$$

which proves the assertion for the one-sided, positive stopping time τ_m^+. The analogue assertions for τ_m^- and τ_m follow in the same manner, where for the result on τ_m^- we also use the fact that $\{W(t) \mid t \geq 0\} \overset{D}{=} \{-W(t) \mid t \geq 0\}$. $\qquad\square$

1.3.2 Asymptotics for unknown in-control parameters

In this section we transfer the results of Theorem 1.3.1 to the more realistic case of unknown in-control parameters.

Theorem 1.3.4. *Let* $\{W(t) \mid t \geq 0\}$ *be a Wiener process. With the notation and assumptions of Section 1.1 and Section 1.2 it holds under the null hypothesis that*

$$\lim_{m \to \infty} P(\hat{\tau}_m^+ = \infty) = P\left(\sup_{t \geq \tilde{G}(1)} W(t)/h(t) \leq 1 \right),$$

$$\lim_{m \to \infty} P(\hat{\tau}_m^- = \infty) = P\left(\sup_{t \geq \tilde{G}(1)} W(t)/h(t) \leq 1 \right),$$

$$\lim_{m \to \infty} P(\hat{\tau}_m = \infty) = P\left(\sup_{t \geq \tilde{G}(1)} |W(t)|/h(t) \leq 1 \right).$$

Proof of Theorem 1.3.4. First of all, we note that we can neglect the estimate $\hat{b}_k$ in our further considerations: Let $\tilde{T}_k := \hat{T}_k \, \hat{b}_k / b$ denote the test statistic with estimated location but non-estimated scale parameter. By the assumption on the convergence of $\hat{b}_k$ (see (1.2.7)) it holds that

$$
\begin{aligned}
\sup_{k=m,\dots,\infty} & \frac{|\hat{T}_k - \tilde{T}_k|}{h(\tilde{G}(k/m))} \\
&\leq \sup_{k=m,\dots,\infty} \frac{|\tilde{T}_k|}{h(\tilde{G}(k/m))} \sup_{k=m,\dots,\infty} |b/\hat{b}_k - 1| \\
&= \sup_{k=m,\dots,\infty} \frac{|\tilde{T}_k|}{h(\tilde{G}(k/m))} \, o_P(1) \\
&= o_P(1),
\end{aligned}
\tag{1.3.7}
$$

where the last equality holds true if we confirm that

$$\sup_{k=m,\dots,\infty} |\tilde{T}_k|/h(\tilde{G}(k/m)) = \mathcal{O}_P(1),$$

which is a consequence of the remaining part of the proof.

We decompose $\tilde{T}_k$ into one part that corresponds to the detector for known in-control parameters and another part that corresponds to the estimation of a:

$$\tilde{T}_k = T_k + \sum_{i=1}^{k} \frac{g(i/m)\,(\hat{a}_k - a)}{b}.
\tag{1.3.8}$$

Replacing G by $\tilde{G}$ in the proof of Theorem 1.3.1 and taking Assumption 1.2.4 into account allows us to approximate T_k as follows:

$$\sup_{k=m,\dots,\infty} \left| \frac{T_k}{h(\tilde{G}(k/m))} - \frac{\int_0^{k/m} g(x)\,dW_m(x)}{h(\tilde{G}(k/m))} \right| = o_P(1).
\tag{1.3.9}$$

In order to approximate the second expression of (1.3.8) in a similar manner we make
use of the invariance principle (1.1.2) to replace the estimate $\hat{a}_k$ by a corresponding
Gaussian one:

$$
\sup_{k=m,\ldots,\infty} \left| \frac{\sum_{i=1}^{k} g(i/m)(\hat{a}_k - a)}{\sqrt{m}\, b\, h(\tilde{G}(k/m))} - \frac{\sum_{i=1}^{k} g(i/m) W_m(k/m)}{k\, h(\tilde{G}(k/m))} \right|
$$

$$
= \sup_{k=m,\ldots,\infty} \left| \frac{\sum_{i=1}^{k} g(i/m)(Z(k) - ka)}{k\, \sqrt{m}\, b\, h(\tilde{G}(k/m))} - \frac{\sum_{i=1}^{k} g(i/m) W(k)}{k\, \sqrt{m}\, h(\tilde{G}(k/m))} \right| \tag{1.3.10}
$$

$$
= \sup_{k=m,\ldots,\infty} \frac{\sum_{i=1}^{k} g(i/m)\, k^{\kappa}}{k\, \sqrt{m}\, h(\tilde{G}(k/m))}\, \mathcal{O}_P(1)
$$

$$
= o_P(1),
$$

where the last equality follows from (1.2.9). We replace the sum by an integral:

$$
\sup_{k=m,\ldots,\infty} \left| \frac{\sum_{i=1}^{k} g(i/m)\, W_m(k/m)}{k\, h(\tilde{G}(k/m))} - \frac{\int_0^{k/m} g(y)\, dy\, W_m(k/m)}{k/m\, h(\tilde{G}(k/m))} \right|
$$

$$
= \sup_{k=m,\ldots,\infty} \frac{\sqrt{\log\log(k/m)k/m}}{k/m\, h(\tilde{G}(k/m))}
$$

$$
\times \left| \frac{1}{m} \sum_{i=1}^{k} g(i/m) - \int_0^{k/m} g(y)\, dy \right| \mathcal{O}_P(1) \tag{1.3.11}
$$

$$
= \sup_{k=m,\ldots,\infty} \frac{\sqrt{\log\log(k/m)}\, g(k/m)}{\sqrt{k/m}\, h(\tilde{G}(k/m))\, m}\, \mathcal{O}_P(1)
$$

$$
= \sup_{k=m,\ldots,\infty} \frac{k^{\kappa}\, g(k/m)}{\sqrt{m}\, h(\tilde{G}(k/m))} \frac{\sqrt{\log\log(k/m)}}{k^{\kappa+\frac{1}{2}}}\, \mathcal{O}_P(1)
$$

$$
= o_P(1),
$$

where the difference between the sum and the integral was approximated in the
obvious manner, i.e.

$$
\left| \frac{1}{m} \sum_{i=1}^{k} g(i/m) - \int_0^{k/m} g(y)\, dy \right|
$$

$$
\leq \frac{1}{m} \sum_{i=1}^{k} g(i/m) - g((i-1)/m)
$$

$$
= g(k/m)/m.
$$

On rewriting

$$
\frac{\int_0^{k/m} g(y)\, dy\, W_m(k/m)}{k/m} \overset{\text{a.s.}}{=} \int_0^{k/m} \frac{\int_0^{k/m} g(y)\, dy}{k/m}\, dW_m(x) \tag{1.3.12}
$$

and combining (1.3.9), (1.3.10), (1.3.11) and (1.3.12) we obtain

$$
\sup_{k=m,\ldots,\infty} \left| \frac{\tilde{T}_k}{h(\tilde{G}(k/m))} - \frac{U(k/m)}{h(\tilde{G}(k/m))} \right| = o_P(1), \tag{1.3.13}
$$

where

$$
U(t) = \begin{cases} \int_0^t \left(g(x) - \dfrac{\int_0^t g(y)\,dy}{t} \right) dW_m(x) & \text{for } t > 0, \\ 0 & \text{for } t = 0. \end{cases} \tag{1.3.14}
$$

The next step is to proceed to a time continuous argument:

$$
\sup_{t \geq m} \left| \frac{U(t/m)}{h(\tilde{G}(t/m))} - \frac{U([t]/m)}{h(\tilde{G}([t]/m))} \right|
$$
$$
\leq \sup_{t \geq m} \frac{|U(t/m)|}{h(\tilde{G}(t/m))} \left| 1 - \frac{h(\tilde{G}(t/m))}{h(\tilde{G}([t]/m))} \right| \tag{1.3.15}
$$
$$
+ \sup_{t \geq m} \frac{|U(t/m) - U([t]/m)|}{h(\tilde{G}([t]/m))},
$$

where in the remaining part of this proof we will see that the first term of (1.3.15) is of order $\mathcal{O}_P(1)$. Thus, by (1.2.11) we have

$$
\sup_{t \geq m} \frac{|U(t/m)|}{h(\tilde{G}(t/m))} \left| 1 - \frac{h(\tilde{G}(t/m))}{h(\tilde{G}([t]/m))} \right| = o_P(1). \tag{1.3.16}
$$

As to the second summand of (1.3.15) we have

$$
\sup_{t \geq m} \frac{|U(t/m) - U([t]/m)|}{h(\tilde{G}([t]/m))}
$$
$$
\leq \sup_{t \geq m} \frac{\left| \int_{[t]/m}^{t/m} g(x)\,dW_m(x) \right|}{h(\tilde{G}([t]/m))}
$$
$$
+ \sup_{t \geq m} \left| \frac{\int_0^{t/m} g(y)\,dy\, W_m(t/m)}{t/m\, h(\tilde{G}([t]/m))} - \frac{\int_0^{[t]/m} g(y)\,dy\, W_m([t]/m)}{[t]/m\, h(\tilde{G}([t]/m))} \right|.
$$

On the one hand, similarly as from (1.3.4) onwards we see that

$$
\sup_{t \geq m} \frac{\left| \int_{[t]/m}^{t/m} g(x)\,dW_m(x) \right|}{h(\tilde{G}([t]/m))} = o_P(1).
$$

On the other hand, by the law of the iterated logarithm, Proposition 1.3.2 and by

(1.2.9) we have

$$\sup_{t\geq m} \left| \frac{\int_0^{t/m} g(y)\, dy\, W_m(t/m)}{t/m\; h(\check{G}([t]/m))} - \frac{\int_0^{[t]/m} g(y)\, dy\, W_m([t]/m)}{[t]/m\; h(\check{G}([t]/m))} \right|$$

$$\leq \sup_{t\geq m} \frac{W_m(t/m)\; \int_{[t]/m}^{t/m} g(y)\, dy}{t/m\; h(\check{G}([t]/m))}$$

$$+ \sup_{t\geq m} \frac{\left|\; \int_0^{[t]/m} g(y)\, dy\, W_m(t/m)\; \right|}{t/m\; h(\check{G}([t]/m))} \left| 1 - \frac{t/m}{[t]/m} \right|$$

$$+ \sup_{t\geq m} \frac{\int_0^{[t]/m} g(y)\, dy\; |W_m(t/m) - W_m([t]/m)|}{[t]/m\; h(\check{G}([t]/m))} \qquad (1.3.17)$$

$$= \sup_{t\geq m} \frac{\sqrt{t/m\; \log\log(t/m)}\; g(t/m)/m}{t/m\; h(\check{G}(t/m))}\; \mathcal{O}_P(1)$$

$$+ \sup_{t\geq m} \frac{g(t/m)\; \sqrt{t/m\; \log\log(t/m)}}{t\; h(\check{G}(t/m))}\; \mathcal{O}_P(1)$$

$$+ \sup_{t\geq m} \frac{g(t/m)\; \sqrt{1/m\; \log(t)}}{h(\check{G}(t/m))}\; \mathcal{O}_P(1)$$

$$= o_P(1).$$

Plugging (1.3.16) and (1.3.17) into (1.3.15) yields

$$\sup_{t\geq m} \left| \frac{U(t/m)}{h(\check{G}(t/m))} - \frac{U([t]/m)}{h(\check{G}([t]/m))} \right| = o_P(1),$$

which again yields in a combination with (1.3.13) that

$$\sup_{k=m,\ldots,\infty} \frac{\tilde{T}_k}{h(\check{G}(k/m))} = \sup_{t\geq m} \frac{U(t/m)}{h(\check{G}(t/m))} + o_P(1). \qquad (1.3.18)$$

Finally, we calculate the covariance of the process $\{U(t)\,|\, t \geq 0\}$, where $U(t)$ is defined in (1.3.14). Let $I(t) := \int_0^t g(x)\, dx/t$ and consider $0 < s \leq t$ first:

$$\mathrm{Cov}(U(s), U(t))$$

$$= \mathrm{Cov}\left(\int_0^s (g(x) - I(s))\, dW_m(x), \int_0^t (g(x) - I(t))\, dW_m(x) \right)$$

$$= \int_0^s (g(x) - I(s))\, (g(x) - I(t))\, dx$$

$$= \int_0^s g^2(x)\, dx - I(s) \int_0^s g(x)\, dx - I(t) \int_0^s g(x)\, dx + s\, I(s)\, I(t)$$

$$= \int_0^s g^2(x)\, dx - I^2(s)\, s - I(t)\, I(s)\, s + I(t)\, I(s)\, s \qquad (1.3.19)$$

$$= \int_0^s g^2(x)\, dx - \frac{1}{s} \left(\int_0^s g(x)\, dx \right)^2$$

$$= \tilde{G}(s)$$

$$= \mathrm{Cov}(W(\check{G}(s)),\, W(\check{G}(t))).$$

For $0 = s \leq t$, on the other hand, we have

$$\text{Cov}(U(s), U(t)) = 0 = \text{Cov}(W(\tilde{G}(s)), W(\tilde{G}(t)))$$

and since $\{U(t) \,|\, t \geq 0\}$ is a Gaussian process it holds true that $\{U(t) \,|\, t \geq 1\} \overset{D}{=} \{W(\tilde{G}(t)) \,|\, t \geq 1\}$. Hence, by $\tilde{G}([1, \infty)) = [\tilde{G}(1), \infty)$ it follows that

$$\sup_{t \geq 1} \frac{U(t)}{h(\tilde{G}(t))} \overset{D}{=} \sup_{t \geq 1} \frac{W(\tilde{G}(t))}{h(\tilde{G}(t))} = \sup_{t \geq \tilde{G}(1)} \frac{W(t)}{h(t)}.$$

Combining this with (1.3.18) completes the proof for the one-sided, positive stopping time τ_m^+. The proof of the corresponding assertions for τ_m^- and τ_m follows by the same arguments. $\qquad\square$

1.3.3   Estimating the scale parameter

In this section we show that the sequential estimate $\hat{b}_k^2$ suggested in Remark 1.2.6 fulfills the convergence stipulated in (1.2.7) under the null hypothesis. In order to preserve the consistency under the alternative, one can estimate b^2 non-sequentially, i.e. use $\hat{b}_k^2 = \hat{b}_m^2$ for all $k \geq m$.

We consider an overlapping batch means estimate which was already studied by Timmermann (2010), hence just the key steps of the proof are presented here.

Lemma 1.3.5. *With the notation and assumptions of Section 1.1 let*

$$\hat{b}_k^2 = \frac{1}{\hat{k}(k - \hat{k} + 1)} \sum_{j=\hat{k}}^{k} \left(Z(j) - Z(j - \hat{k}) - \hat{k}\, Z(k)/k \right)^2, \tag{1.3.20}$$

where $\hat{k} = \hat{k}_k$ is some integer sequence such that $\hat{k} \to \infty$ but $\hat{k}/k \to 0$ as $k \to \infty$. Then it holds under the null hypothesis that, as $m \to \infty$,

$$\sup_{k=m,\dots,\infty} |\,\hat{b}_k^2 - b^2\,| = \sup_{k=m,\dots,\infty} \left(k^\kappa \sqrt{\log(k/\hat{k})}/\sqrt{\hat{k}} + \sqrt{\log\log(k/\hat{k})}\,\hat{k}/k \right) O_P(1).$$

Remark 1.3.6. *On setting $\hat{k} = k^q$ for some $2\kappa < q < 1$ we obtain the desired consistency of the estimate, that is under the null hypothesis that, as $m \to \infty$,*

$$\sup_{k=m,\dots,\infty} |\,\hat{b}_k^2 - b^2| = o_P(1).$$

Proof of Lemma 1.3.5. The idea of the proof is to decompose the overlapping batch means estimate $\hat{b}_k^2$ into $\hat{k}$ non-overlapping batch means estimates

$$\breve{b}_{k,r}^2 = \frac{1}{\hat{k}\,[k/\hat{k}]} \sum_{j=1}^{[k/\hat{k}]} \left(Z(j, \hat{k}, r, k, k) \right)^2 \, 1_{[0,k]}(j\hat{k} + r), \tag{1.3.21}$$

where $r = 0, \dots, \hat{k} - 1$ and

$$Z(j, \hat{k}, r, l, k) := Z(j\,\hat{k} + r) - Z((j-1)\hat{k} + r) - \hat{k}Z(l)/k.$$

Note that

$$\hat{b}_k^2 = \frac{\hat{k}\,[k/\hat{k}]}{(k - \hat{k} + 1)\,\hat{k}} \sum_{r=0}^{k-1} \breve{b}_{k,r}^2,$$

hence, if we show that

$$\sup_{k=m,\dots,\infty} \max_{r=0,\dots,\hat{k}-1} |\,\breve{b}_{k,r}^2 - b^2\,|$$

$$= \sup_{k=m,\dots,\infty} \left(k^\kappa \sqrt{\log(k/\hat{k})}/\sqrt{\hat{k}} + \sqrt{\log\log(k)}\,\hat{k}/k \right) O_P(1),$$

the assertion follows by

$$\sup_{k=m,\ldots,\infty} \frac{\hat{k}\,[k/\hat{k}]}{k-\hat{k}+1} = \mathcal{O}(1).$$

Denoting

$$W(j,\hat{k},r,l,k) := W(j\,\hat{k}+r) - W((j-1)\hat{k}+r) - \hat{k}W(l)/k,$$

we have

$$\left| \check{b}_{k,r}^2 - \frac{1}{\hat{k}\,[k/\hat{k}]} \sum_{j=1}^{[k/\hat{k}]} b^2 \left(W(j,\hat{k},r,k,k) \right)^2 1_{[0,k]}(j\hat{k}+r) \right|$$
$$\leq \max_{j=1,\ldots,[k/\hat{k}]} \left(\frac{|\,Z(j,\hat{k},r,k,k) - b\,W(j,\hat{k},r,k,k)\,|}{\sqrt{k}} \right.$$
$$\left. \times \; \frac{|\,Z(j,\hat{k},r,k,k) - b\,W(j,\hat{k},r,k,k) + 2\,W(j,\hat{k},r,k,k)\,|}{\sqrt{k}} \right).$$

Applying the invariance principle (1.1.2) yields uniformly for $k = m,\ldots,\infty$, $r = 0,\ldots,\hat{k}-1$ and $j = 1,\ldots,[k/\hat{k}]$ that

$$\frac{|\,Z(j,\hat{k},r,k,k) - b\,W(j,\hat{k},r,k,k)\,|}{\sqrt{k}} = \sup_{k=m,\ldots,\infty} k^{\kappa}/\sqrt{k}\,\mathcal{O}_P(1).$$

Further, by Proposition 1.3.2 and the law of the iterated logarithm we have uniformly for $k = m,\ldots,\infty$, $r = 0,\ldots,\hat{k}-1$ and $j = 1,\ldots,[k/\hat{k}]$ that

$$\frac{|\,W(j,\hat{k},r,l,k)\,|}{\sqrt{k}} = \sup_{k=m,\ldots,\infty} \sqrt{\log(k/\hat{k})}\,\mathcal{O}_P(1),$$

where a combination of the two convergence rates from above yields the first rate in (1.3.20). Finally, by the law of the iterated logarithm, making use of the fact that the increments $(W(i) - W(i-1))^2$ are independently, identically distributed with mean 1 we have uniformly for $k = m,\ldots,\infty$, $r = 0,\ldots,\hat{k}-1$ and $j = 1,\ldots,[k/\hat{k}]$ that

$$\frac{1}{\hat{k}\,[k/\hat{k}]} \sum_{j=1}^{[k/\hat{k}]} \left(\left(W(j,\hat{k},r,k,k) \right)^2 - \hat{k} \right) 1_{[0,k]}(j\hat{k}+r)$$
$$\overset{D}{=} \frac{1}{[k/\hat{k}]} \sum_{j=1}^{[k/\hat{k}]} \left(\left(W(j,1,r/\hat{k},k,k/\hat{k}) \right)^2 - 1 \right) 1_{[0,k]}(j\hat{k}+r)$$
$$\overset{\text{a.s.}}{=} \mathcal{O}(1) \sup_{k=m,\ldots,\infty} \sqrt{\log\log(k/\hat{k})\hat{k}/k},$$

which completes the proof. $\qquad\square$

1.4   Consistency of the testing procedures

The monitoring procedure is said to be consistent, if the probability of a correct detection of a change (i.e. a stopping of the monitoring procedure at some time point $N < \infty$) converges towards 1 as $m \to \infty$. This is shown in Theorem 1.4.1 and Theorem 1.4.3 of Section 1.4.1 or Section 1.4.2, respectively, under rather general assumptions on the type of change, i.e. under general growth conditions which depend on the type of gradual change $\Delta_{m,\gamma}$, the weight function g and the threshold function h. In Section 1.4.3 we analyze the growth conditions for the type of change introduced in Example 1.1.1 and Example 1.1.2 and the weight and threshold functions suggested in Remark 1.2.2 and Remark 1.2.5.

Note that throughout this section the assumptions on the process $Y^*(t - T^*)$ can be weakened, namely we can replace Assumption (1.1.3) by

$$\sup_{T^* \leq t \leq T} |Y^*(t - T^*)| = \mathcal{O}_P(\sqrt{T}) \tag{1.4.1}$$

as $T \to \infty$. This is of course the case if the invariance principle of (1.1.3) holds true.

1.4.1   Consistency for known in-control parameters

Theorem 1.4.1. *With the notation and assumptions of Section 1.1 and Section 1.2 it holds that*

$$\text{under } H_1^+ \quad \lim_{m \to \infty} P\big(\tau_m^+ \leq N\big) = 1,$$

$$\text{under } H_1^- \quad \lim_{m \to \infty} P\big(\tau_m^- \leq N\big) = 1,$$

$$\text{under } H_1 \quad \lim_{m \to \infty} P\big(\tau_m \leq N\big) = 1,$$

where $N = N_m > T^*$ is an arbitrary integer sequence fulfilling the following growth conditions:

$$\frac{g(N/m)\,\sqrt{N/m}}{h(G(N/m))} = \mathcal{O}(1), \tag{1.4.2}$$

but

$$\frac{|\Lambda_N|}{h(G(N/m))} \to \infty, \tag{1.4.3}$$

where

$$\Lambda_k := \sum_{i=K^*+1}^{k} \frac{g(i/m)\,(\Delta_{m,\gamma}(i - T^*) - \Delta_{m,\gamma}(i - 1 - T^*))}{\sqrt{m}\,b}$$

denotes the deterministic perturbation of the detector T_k and $K^* := [T^*]$ denotes the last integer time point before the change point.

Remark 1.4.2. *Possible choices of weight and threshold functions are given in Remark 1.2.2. If the change is of the type as introduced in Example 1.1.1 and Example 1.1.2, i.e.*

$$\Delta_{m,\gamma}(i - T^*) - \Delta_{m,\gamma}(i - 1 - T^*) = \delta\big((i - T^*)/m\big)_{+}^{\gamma},$$

where $\delta \neq 0$ shall be constant and $\gamma > 0$, a possible choice of N is $N = [\rho T^*]$ where $1 < \rho < (\lambda + 1)^{1/\lambda}$. If we choose $h^{(1)}$, Assumptions (1.4.2) and (1.4.3) are fulfilled for any $\gamma \geq \lambda > 0$ and if we choose $h^{(2)}$ or $h^{(3)}$, Assumptions (1.4.2) and (1.4.3) are fulfilled for arbitrary $\gamma, \lambda > 0$, which is verified in Section 1.7 below.

Proof of Theorem 1.4.1. As stated above, let $N = N_m > T^*$ be an integer sequence depending of m. Since T^* will not be integer in general we consider $K^* = [T^*]$, so $Z(K^* + 1)$ is the first observation we make after the change point. By Theorem 1.3.1 the observations before $K^* + 1$ are bounded in the following sense:

$$\begin{aligned}
T_N &= \sum_{i=K^*+1}^{N} \frac{g(i/m)\,(Z_i - a)}{\sqrt{m}\,b} + \sum_{i=1}^{K^*} \frac{g(i/m)\,(Z_i - a)}{\sqrt{m}\,b} \\
&= \sum_{i=K^*+1}^{N} \frac{g(i/m)\,(Z_i - a)}{\sqrt{m}\,b} + \mathcal{O}_P(h(G(K^*/m))).
\end{aligned}$$

We decompose the remaining term into its stochastic part and its deterministic part and show that under the growth conditions (1.4.2) and (1.4.3) the latter one is dominating and tends to infinity. Via Remark 1.3.3 we have

$$\begin{aligned}
\sum_{i=K^*+1}^{N} \frac{g(i/m)(Z_i - a)}{\sqrt{m}\,b} &= \frac{g(N/m)(Z(N) - aN)}{\sqrt{m}\,b} \\
&\quad - \sum_{i=K^*+1}^{N-1} \frac{(g((i+1)/m) - g(i/m))(Z(i) - ai)}{\sqrt{m}\,b} \\
&\quad - \frac{g((K^* + 1)/m)\,(Z(K^*) - aK^*)}{\sqrt{m}\,b} \\
&= \frac{g(N/m)}{\sqrt{m}} \left(Y(T^*) + \frac{b^*\,Y^*(N - T^*)}{b} + \frac{\Delta_{m,\gamma}(N - T^*)}{b} \right) \\
&\quad - \sum_{i=K^*+1}^{N-1} \frac{g((i+1)/m) - g(i/m)}{\sqrt{m}} \\
&\qquad \times \left(Y(T^*) + \frac{b^*\,Y^*(i - T^*)}{b} + \frac{\Delta_{m,\gamma}(i - T^*)}{b} \right) \\
&\quad - \frac{g((K^* + 1)/m)\,Y(K^*)}{\sqrt{m}}
\end{aligned}$$

$$
\begin{aligned}
= \; & \sum_{i=K^*+1}^{N} \frac{g(i/m)\left(\Delta_{m,\gamma}(i-T^*) - \Delta_{m,\gamma}(i-1-T^*)\right)}{\sqrt{m}\, b} \\
& + \frac{g(N/m)\, b^*\, Y^*(N-T^*)}{\sqrt{m}\, b} \\
& - \sum_{i=K^*+1}^{N-1} \frac{\left(g((i+1)/m) - g(i/m)\right) b^*\, Y^*(i-T^*)}{\sqrt{m}\, b} \\
& + \frac{g((K^*+1)/m)\left(Y(T^*) - Y(K^*)\right)}{\sqrt{m}} \\
= \; & \Lambda_N + \mathcal{O}_P\!\left(\sqrt{N/m}\; g(N/m)\right).
\end{aligned}
$$

Combining this with (1.4.2) and (1.4.3) yields the assertion for the one-sided stopping time τ_m^+ :

$$
\begin{aligned}
P\!\left(\tau_m^+ < \infty\right) & \\
\geq \; & P\!\left(T_N/h(G(N/m)) > 1\right) \\
= \; & P\Big(\Lambda_N/h(G(N/m)) \\
& \quad + \mathcal{O}_P\!\left(\sqrt{N/m}\; g(N/m)/h(G(N/m))\right) + \mathcal{O}_P(1) > 1\Big) \\
\to \; & 1.
\end{aligned}
$$

The analogue assertions for τ_m^- and τ_m follow in the same manner. □

1.4.2 Consistency for unknown in-control parameters

Theorem 1.4.3. *With the notation and assumptions of Section 1.1 and Section 1.2 it holds that*

$$\text{under } H_1^+ \quad \lim_{m\to\infty} P\big(\hat{\tau}_m^+ \leq N\big) = 1,$$

$$\text{under } H_1^- \quad \lim_{m\to\infty} P\big(\hat{\tau}_m^- \leq N\big) = 1,$$

$$\text{under } H_1 \quad \lim_{m\to\infty} P\big(\hat{\tau}_m \leq N\big) = 1,$$

where $N = N_m > T^*$ is an arbitrary integer sequence fulfilling the following growth conditions:

$$\frac{g(N/m)\,\sqrt{N/m}}{\hat{b}_N\,h(\tilde{G}(N/m))} = \mathcal{O}_P(1), \tag{1.4.4}$$

but

$$\frac{|\hat{\Lambda}_N|}{\hat{b}_N\,h(\tilde{G}(N/m))} \xrightarrow{P} \infty, \tag{1.4.5}$$

with $K^* = [T^*]$, $\hat{b}_k$ being an estimate for which $\max_{k=m,\dots,K^*}|\hat{b}_k^2 - b^2| = o_P(1)$ holds true and

$$\hat{\Lambda}_k = \sum_{i=K^*+1}^{k} \frac{\big(g(i/m) - \sum_{j=1}^{k} g(j/m)/k\big)\big(\Delta_{m,\gamma}(i - T^*) - \Delta_{m,\gamma}(i - 1 - T^*)\big)}{\sqrt{m}\,b}$$

being the deterministic perturbation of T_k.

Remark 1.4.4. *By estimating b non-sequentially, i.e. not with every newly made observation but rather by only using the data obtained during the training period, one can preserve the consistency of the estimate even if a change occurred. So, if we choose $\hat{b}_N = \hat{b}_m = b + o_P(1)$, (see e.g. Remark 1.2.6 and Section 1.3.3), the estimate $\hat{b}_N$ in (1.4.4) and (1.4.5) can be dropped.*

Remark 1.4.5. *Possible choices of weight and threshold functions are given in Remark 1.2.5. If the change is of the type as introduced in Example 1.1.1 and Example 1.1.2, i.e.*

$$\Delta_{m,\gamma}(i - T^*) - \Delta_{m,\gamma}(i - 1 - T^*) = \delta\big((i - T^*)/m\big)_+^\gamma,$$

where $\delta \neq 0$ shall be constant and $\gamma > 0$, a possible choice of N is $N = [\rho T^*]$ where $1 < \rho < (\lambda + 1)^{1/\lambda}$. If we choose $h^{(1)}$, Assumptions (1.4.4) and (1.4.5) are fulfilled for any $\gamma \geq \lambda > 0$ and if we choose $h^{(2)}$ or $h^{(3)}$, Assumptions (1.4.4) and (1.4.5) are fulfilled for arbitrary $\gamma, \lambda > 0$, which is verified in Section 1.7 below.

Proof of Theorem 1.4.3. Similar as in the proof of Theorem 1.4.1 we consider the statistic for $k = N$ and decompose the detector into its stochastic and its deterministic parts, which yields

$$\hat{T}_N = \hat{\Lambda}_N/\hat{b}_N + \mathcal{O}_P\big(\sqrt{N/m}\; g(N/m)/\hat{b}_N\big) + \mathcal{O}_P\big(h(\tilde{G}(K^*/m))\big).$$

Now the assertions follow by the growth conditions (1.4.4) and (1.4.5).   □

1.4.3 Local alternatives

In this section we focus on the type of change introduced in Example 1.1.1 and Example 1.1.2, i.e. on

$$\Delta_{m,\gamma}(t) = \delta_m \sum_{i=1}^{[t]} \left((i - T^*)/m \right)_+^{\gamma},$$

where we particularly consider local alternatives, that is $\delta_m \to 0$ as $m \to \infty$. Our objective is to find an upper bound for the time the test takes to detect a change depending on the rate of convergence of δ_m and on the time when the change occurred.

For the sake of simplicity we assume the change point T^* to be integer, however, the same asymptotics hold true if this is not the case. Also, we focus on the more realistic setting of unknown in-control parameters a and b. For known in-control parameters the results can be obtained in the same manner.

We consider the weight function suggested in Remark 1.2.5, namely $g(t) = t^\lambda$ for some $\lambda > 0$, which implies $\tilde{G}(t) = \tilde{\lambda} t^{2\lambda+1}$ where $\tilde{\lambda} = \lambda^2/((2\lambda+1)(\lambda+1)^2)$. We aim to find a "small" N such that $P(\hat{\tau} \leq N) \to 1$, where $\hat{\tau}$ represents either of the stopping times $\hat{\tau}_m^+, \hat{\tau}_m^-$ or $\hat{\tau}_m$. By the calculations given in Section 1.7 (assuming that we have an estimate $\hat{b}_N = b + o_P(1)$ and that $N - T^* \to \infty$) it is sufficient to find a sequence $N = N_m$ such that

$$F_1 := \frac{|\delta_m| (N - T^*)^{1+\gamma} (N/m)^\lambda}{m^{1/2+\gamma} h(\tilde{G}(N/m))} \simeq \frac{|\delta_m| R^{1+\gamma} (N/m)^\lambda}{m^{1/2+\gamma} h(\tilde{G}(N/m))} \to \infty,$$

where $R = N - T^*$ is an upper bound for the time the procedures take to detect a change in the sense that $P(\hat{\tau} - T^* \leq R) \to 1$. The choice of N, however, depends on whether R or T^* is dominant, hence we distinguish three cases:

$\underline{R/T^* \to 0:}$ If

$$\frac{1}{T^*} \frac{m^{1/2+\gamma} h(\tilde{G}(T^*/m))}{(T^*/m)^\lambda |\delta_m|} \to 0, \tag{1.4.6}$$

one may choose

$$R = \left(\frac{m^{1/2+\gamma} h(\tilde{G}(T^*/m))}{(T^*/m)^\lambda |\delta_m|} \right)^{1/(1+\gamma)} r(m),$$

where $r(m) \to \infty$ such that $R \to \infty$, but $R/T^* \to 0$. Now, plugging R into F_1 yields $F_1 \simeq r(m)^{1+\gamma} \to \infty$.

$\underline{T^* \simeq R:}$ If (1.4.6) does not hold true, but

$$\frac{|\delta_m| (T^*/m)^{1+\lambda+\gamma} \sqrt{m}}{h(\tilde{G}(T^*/m))} \to \infty \tag{1.4.7}$$

does, one can choose $R = T^*$. In this case we have

$$F_1 \simeq \frac{|\delta_m|\,(T^*/m)^{1+\lambda+\gamma}\,\sqrt{m}}{h(\breve{G}(T^*/m))} \to \infty.$$

$\underline{T^*/R \to 0}$: If neither (1.4.6) nor (1.4.7) holds true, one has to choose R large enough such that, for one thing, $R/T^* \to \infty$ and, for another thing,

$$F_1 \simeq \frac{|\delta_m|\,(R/m)^{1+\lambda+\gamma}\,\sqrt{m}}{h(\breve{G}(R/m))} \to \infty.$$

Choosing $h = h^{(1)}$ of Remark 1.2.5 and $\gamma > \lambda > 0$ a possible choice is

$$R = \max\left\{\left(|\delta_m|\sqrt{m}\right)^{-1/(\gamma-\lambda)} m,\, T^*\right\} r(m),$$

and choosing $h = h^{(2)}$ or $h = h^{(3)}$, respectively, and if $|\delta_m|\sqrt{m} \not\to 0$, a possible choice is

$$R = \max\left\{m\,(|\delta_m|\sqrt{m})^{-1/(1+\gamma)},\, T^{*1+\epsilon}\right\} r(m)$$

for some $\epsilon > 0$ and if $|\delta_m|\sqrt{m} \to 0$, a possible choice is

$$R = \max\left\{m\,(|\delta_m|\sqrt{m})^{-1/(1+\gamma)-\epsilon},\, T^{*1+\epsilon}\right\} r(m)$$

for some $\epsilon > 0$, where in all three cases $r(m) \to \infty$ as $m \to \infty$.

1.5 Asymptotic normality of the delay times

The results of the previous section state that under quite mild assumptions a change is detected with probability tending to one. However, Theorem 1.4.1 and Theorem 1.4.3 do not give us much information about when a (detected) change took place. In this section we will prove limit distributions for delay times, i.e. the time lag between the change point and its detection. It turns out (under the assumptions below) that the suitably standardized delay times are asymptotically normally distributed and the results can be used to establish asymptotic confidence intervals for the change point. However, these more sophisticated results also require a more narrow setting compared to the one in Section 1.4. Particularly, we restrict the considerations to changes which fulfill

$$\Delta_{m,\gamma}(i - T^*) - \Delta_{m,\gamma}(i - 1 - T^*) = \delta\big((i - T^*)/m\big)_+^{\gamma}, \tag{1.5.1}$$

where $\delta \neq 0$ is constant (see also Example 1.1.1 and Example 1.1.2). The techniques of the proofs of Theorem 1.5.1 and Theorem 1.5.7 also require us to know the "direction" of the change, that is whether δ is positive or negative, hence the asymptotic distributions can only be shown for the one-sided stopping times τ_m^+, τ_m^-, $\hat{\tau}_m^+$ and $\hat{\tau}_m^-$. Nevertheless, the results for the one-sided stopping times can be used to construct asymptotic confidence intervals under the two sided alternative, as well.

A key assumption in this section is an early-change scenario (see Assumptions (1.5.3) and (1.5.19)) which states that the change occurs soon after the training period. Such early-change settings are common assumptions if one wants to prove asymptotic normality of the delay times, cf. e.g. Aue and Horváth (2004) and Gut and Steinebach (2009).

We stick to the pattern of showing the results for known in-control parameters a and b first (Section 1.5.1), and then carrying those results over to the case of unknown parameters (Section 1.5.2). Moreover, since we are working under the alternative the standardization for the asymptotic distribution also depends on γ and δ, therefore we introduce a consistent estimate for δ. The slope parameter γ is assumed to be known.

1.5.1 Asymptotic normality of the delay times for known in-control parameters

Theorem 1.5.1. *With the notation and assumptions of Section 1.1 and Section 1.2 as well as $h(G(t))$ being continuous at $t = 1$, (1.5.1) holding true with δ being constant,*

$$\sup_{1 \leq \xi \leq H/m} |g'(\xi)| = \mathcal{O}(1) \tag{1.5.2}$$

(where H is defined in (1.5.5) below) and the early-change assumption

$$T^* = m + o(m) \tag{1.5.3}$$

it holds for all $x \geq -h(G(1))/\sqrt{G(1)}$ *under* H_1^+ *that*

$$\lim_{m \to \infty} P\left(\frac{\left(\tau_m^+ - T^*\right)_+^{1+\gamma} g(1) \, \delta}{m^{1/2+\gamma} \, (1+\gamma) \, b \, \sqrt{G(1)}} - \frac{h(G(1))}{\sqrt{G(1)}} \leq x \right) = \phi(x),$$

and under H_1^- *that*

$$\lim_{m \to \infty} P\left(\frac{\left(\tau_m^- - T^*\right)_+^{1+\gamma} g(1) \, |\delta|}{m^{1/2+\gamma} \, (1+\gamma) \, b \, \sqrt{G(1)}} - \frac{h(G(1))}{\sqrt{G(1)}} \leq x \right) = \phi(x).$$

The rather longish proof of Theorem 1.5.1 is given further below.

By Slutsky's lemma the convergences above still hold true if we replace δ by a consistent estimate $\hat{\delta}$, where a possible choice is suggested in Lemma 1.5.2. Note that in order to obtain consistency for $\hat{\delta}$ we need to have a sufficient amount of observations made after the change point which might require to let the process $Z(t)$ run on (in its perturbed form) in order to determine at which time point the change has occurred. In the following situation we obtain a consistent estimate:

Lemma 1.5.2. *With the notation and assumptions of Section 1.1 let*

$$\hat{\delta}_{m,k} = \frac{\sum_{i=1}^{k} ((i-m)/m)_+^{\gamma} \, (Z_i - a)}{\sum_{i=1}^{k} ((i-m)/m)_+^{2\gamma}}.$$

If for some $\zeta > 1$ *it holds that*

$$T^* - m = o(m^{\zeta}),$$

then we have under either of the alternatives H_1^+, H_1^- *or* H_1 *that*

$$\hat{\delta}_{m,m+m^{\zeta}} = \delta + o_P(1).$$

In Section 1.5.2 we give a proof for a slightly more general version of Lemma 1.5.2 (i.e. for a unknown), hence the proof is omitted here. Combining Theorem 1.5.1 and Lemma 1.5.2 gives us the following asymptotic confidence intervals for the change point:

Corollary 1.5.3. *Let the assumptions of Theorem 1.5.1 hold true. On setting*

$$q(\chi_i) = \left(\left(\frac{h(G(1))}{\sqrt{G(1)}} + \chi_i \right) \frac{(1+\gamma) \, b \, \sqrt{G(1)} \, m^{1/2+\gamma}}{g(1) \, |\hat{\delta}|} \right)^{\frac{1}{1+\gamma}},$$

where $\hat{\delta} \xrightarrow{P} \delta$ *is a consistent estimate (e.g.* $\hat{\delta}_{m,m+m^{\zeta}}$ *from Lemma 1.5.2 for some* $\zeta > 1$ *) and*

$$\chi_i = \begin{cases} \phi^{-1}\left(1 - \tilde{\alpha} + \phi(-h(G(1))/\sqrt{G(1)})\right) & \text{if } i = 1, \\ \phi^{-1}\left(1 - \tilde{\alpha} + 2\phi(-h(G(1))/\sqrt{G(1)})\right) & \text{if } i = 2, \end{cases}$$

with

$$\tilde{\alpha} \in \begin{cases} \left(\phi(-h(G(1))/\sqrt{G(1)}),\, 1\right) & for \ \ i = 1, \\ \left(2\,\phi(-h(G(1))/\sqrt{G(1)}),\, 1\right) & for \ \ i = 2, \end{cases} \tag{1.5.4}$$

we obtain for τ_m^+ *and* τ_m^- *as in Section 1.2.1 and* $\tilde{\tau}_m := \min\{\tau_m^+, \tau_m^-\}$ *that*

$$P\big(\tau_m^+ - q(\chi_1) \leq T^* < \tau_m^+\big) = 1 - \tilde{\alpha} + o(1),$$
$$P\big(\tau_m^- - q(\chi_1) \leq T^* < \tau_m^-\big) = 1 - \tilde{\alpha} + o(1),$$
$$P\big(\tilde{\tau}_m - q(\chi_2) \leq T^* < \tilde{\tau}_m\big) \geq 1 - \tilde{\alpha} + o(1)$$

under H_1^+, H_1^- *or* H_1, *respectively.*

Remark 1.5.4. *Note that in Remark 1.2.2 we suggest different threshold functions for the one-sided and two-sided stopping times, corresponding to the known (limit) distributions of Remark 1.2.3. However, if one applies the same threshold function for the one-sided and two-sided stopping times, it holds that* $\tau_m = \tilde{\tau}_m$.

The proof of Corollary 1.5.3 is given at the very end of this section. First, we give the proof of the main theorem of this section, which is Theorem 1.5.1. For a better readability the proof is subdivided into three steps and the following notation is introduced: For $x \geq -h(G(1))/\sqrt{G(1)}$ we set

$$H := K^* + \left[\nu\, m^{\frac{1/2+\gamma}{1+\gamma}}\right], \tag{1.5.5}$$

where

$$K^* := [T^*], \tag{1.5.6}$$

$$\nu := \left(\left(1 + x\,\frac{\sqrt{G(1)}}{h(G(1))}\right) \frac{(1+\gamma)\,b\,h(G(1))}{g(1)\,|\delta|}\right)^{\frac{1}{1+\gamma}}. \tag{1.5.7}$$

Further, let

$$\Lambda_k := \frac{\sum_{i=K^*+1}^{k} g(i/m)\, \delta\,((i - T^*)/m)^{\gamma}}{\sqrt{m}\, b} \tag{1.5.8}$$

denote the perturbation of the detectors. We state and prove two technical lemmata which will be used in the proof of the main theorem. The first lemma covers the convergence of some deterministic terms.

Lemma 1.5.5. *Under the assumptions of Theorem 1.5.1 it holds for all* $x \geq -h(G(1))/\sqrt{G(1)}$ *that, as* $m \to \infty$,

$$\frac{H - T^*}{m} \to 0, \tag{1.5.9}$$

$$H \sim K^*, \tag{1.5.10}$$

$$\frac{\operatorname{sgn}(\delta)\, \Lambda_H}{h(G(H/m))} \to 1 + x\,\frac{\sqrt{G(1)}}{h(G(1))}, \tag{1.5.11}$$

$$\max_{k=K^*+1,\dots,H} \frac{\operatorname{sgn}(\delta)\, \Lambda_k}{h(G(k/m))} \to 1 + x\,\frac{\sqrt{G(1)}}{h(G(1))}, \tag{1.5.12}$$

where H, K^* *and* Λ_k *are defined in (1.5.5), (1.5.6) and (1.5.8).*

Proof of Lemma 1.5.5. The first two assertions are immediate consequences of the definition of H and of the early-change assumption (1.5.3). On showing (1.5.11) we first note that the early-change assumption (1.5.3) allows us to approximate the argument of the weight function and the threshold function in the considered expression by 1:

$$\left| \frac{\Lambda_H}{h(G(H/m))} - \frac{\sum_{i=K^*+1}^{H} g(1)\,\delta\left((i-T^*)/m\right)^\gamma}{\sqrt{m}\,b\,h(G(1))} \right|$$

$$\leq \frac{\sum_{i=K^*+1}^{H} |\delta|\left((i-T^*)/m\right)^\gamma}{\sqrt{m}\,b} \max_{j=K^*+1,\dots,H} \left| \frac{g(j/m)}{h(G(H/m))} - \frac{g(1)}{h(G(1))} \right|.$$

On the one hand, by the definition of H it holds that

$$\frac{\sum_{i=K^*+1}^{H}((i-T^*)/m)^\gamma}{\sqrt{m}}$$

$$\leq \left(\frac{H-T^*}{m} \right)^\gamma \frac{(H-K^*)}{\sqrt{m}}$$

$$= \left(m^{(1/2+\gamma)/(1+\gamma)-1} \right)^{1+\gamma} \sqrt{m}\, \mathcal{O}(1)$$

$$= \mathcal{O}(1)$$

and, on the other hand, by the continuity of the weight function and of the threshold function it holds that

$$\max_{j=K^*+1,\dots,H} \left| \frac{g(j/m)}{h(G(H/m))} - \frac{g(1)}{h(G(1))} \right|$$

$$\leq \max_{j=K^*+1,\dots,H} \left| \frac{g(j/m)-g(1)}{h(G(H/m))} \right| + \left| \frac{g(1)}{h(G(H/m))} - \frac{g(1)}{h(G(1))} \right|$$

$$= \frac{g(H/m)-g(1)}{h(G(H/m))} + o(1)$$

$$= o(1).$$

Thus it is sufficient to consider $\sum_{i=K^*+1}^{H}((i-T^*)/m)^\gamma/\sqrt{m}$. In a next step, we replace the usually non-integer change point T^* by K^*, which is the last integer before the change point. By $0 \leq T^* - K^* \leq 1$ we have

$$\left| \sum_{i=K^*+1}^{H} \left(\frac{i-K^*}{m} \right)^\gamma - \sum_{i=K^*+1}^{H} \left(\frac{i-T^*}{m} \right)^\gamma \right|$$

$$= \sum_{i=1}^{H-K^*} \left(\frac{i}{m} \right)^\gamma - \left(\frac{i-(T^*-K^*)}{m} \right)^\gamma$$

$$\leq \sum_{i=1}^{H-K^*} \left(\frac{i}{m} \right)^\gamma - \left(\frac{i-1}{m} \right)^\gamma$$

$$= \left(\frac{H-K^*}{m} \right)^\gamma$$

$$= o(1).$$

Now we can show that $\sum_{i=K^*+1}^{H}((i-K^*)/m)^\gamma/\sqrt{m}$ converges towards $\nu^{1+\gamma}/(1+\gamma)$, where ν is defined in (1.5.7):

$$
\frac{1}{\sqrt{m}}\sum_{i=K^*+1}^{H}\left(\frac{i-K^*}{m}\right)^\gamma
$$
$$
= \frac{(H-K^*)^{1+\gamma}}{m^{\gamma+1/2}}\sum_{i=1}^{H-K^*}\left(\frac{i}{H-K^*}\right)^\gamma\frac{1}{H-K^*}
$$
$$
= \frac{(H-K^*)^{1+\gamma}}{m^{\gamma+1/2}}\left(\frac{1}{1+\gamma}+o(1)\right)
$$
$$
= \frac{\nu^{1+\gamma}}{1+\gamma}+o(1).
$$

Combining the assertions above we have

$$
\frac{\mathrm{sgn}(\delta)\,\Lambda_H}{h(G(H/m))} = \frac{\mathrm{sgn}(\delta)\,g(1)\,\delta\,\nu^{1+\gamma}}{b\,h(G(1))\,(1+\gamma)}+o(1)
$$
$$
= 1+x\,\sqrt{G(1)}/h(G(1))+o(1),
$$

which proves assertion (1.5.11). On proving (1.5.12) note that by the fact that $\mathrm{sgn}(\delta)\,\Lambda_k$ and $h(G(k/m))$ are increasing in k we have

$$
\frac{\mathrm{sgn}(\delta)\,\Lambda_H}{h(G(H/m))} \leq \max_{k=K^*+1,\dots,H}\frac{\mathrm{sgn}(\delta)\,\Lambda_k}{h(G(k/m))} \leq \frac{\mathrm{sgn}(\delta)\,\Lambda_H}{h(G(H/m))}\frac{h(G(H/m))}{h(G(K^*/m))},
$$

hence (1.5.12) follows by the continuity of $h(G(t))$ at $t=1$. $\square$

The next lemma allows us to replace the current detector (at time point $k>T^*$) by the last detector before the change point and the deterministic perturbation of T_k.

Lemma 1.5.6. *Under the assumptions of Theorem 1.5.1 it holds that, as $m\to\infty$,*

$$
\max_{k=m,\dots,H}\frac{|\,T_k-T_{K^*}-\Lambda_k\,|}{h(G(k/m))} = o_P(1).
$$

Proof of Lemma 1.5.6. We demonstrate the proof for $k=K^*+1,\dots,H$. For $k=m,\dots,K^*$ the assertion follows by the same argument, yet in a more simple manner. By the definition of $Z(t)$ we have for $k=K^*+1,\dots,H$

$$
\frac{|\,T_k-T_{K^*}-\Lambda_k\,|}{h(G(k/m))}
$$
$$
\leq \sum_{i=K^*+2}^{k}\frac{g(i/m)\,|\,Y^*(i-T^*)-Y^*(i-1-T^*)\,|}{\sqrt{m}\,b\,h(G(k/m))}
$$
$$
+\frac{g((K^*+1)/m)\,|\,Y^*(K^*+1-T^*)\,|}{\sqrt{m}\,b\,h(G(k/m))}
$$
$$
+\frac{g((K^*+1)/m)\,|\,Y(T^*)-Y(K^*)\,|}{\sqrt{m}\,b\,h(G(k/m))}.
$$

Hence, by the invariance principles (1.1.2) and (1.1.3) it holds that

$$
\max_{k=K^*+1,\dots,H} \frac{|T_k - T_{K^*} - \Lambda_k|}{h(G(k/m))}
$$

$$
\leq \max_{k=K^*+1,\dots,H} \sum_{i=K^*+2}^{k} \frac{g(i/m)\,|W^*(i-T^*) - W^*(i-1-T^*)|}{\sqrt{m}\, b\, h(G(k/m))}
$$

$$
+ \max_{k=K^*+1,\dots,H} \frac{g((K^*+1)/m)\,|W^*(K^*+1-T^*)|}{\sqrt{m}\, b\, h(G(k/m))}
$$

$$
+ \max_{k=K^*+1,\dots,H} \frac{g((K^*+1)/m)\,|W(T^*) - W(K^*)|}{\sqrt{m}\, b\, h(G(k/m))},
$$

$$
+ \max_{k=K^*+1,\dots,H} \frac{g(k/m)\, H^\kappa}{\sqrt{m}\, h(G(k/m))}\, \mathcal{O}_P(1),
$$

where by (1.2.3), (1.5.3) and (1.5.10) the last term is of order $o_P(1)$. A further approximation by Remark 1.3.3 yields

$$
\sum_{i=K^*+2}^{k} g(i/m)\,|W^*(i-T^*) - W^*(i-1-T^*)|
$$

$$
\leq 4\,g(k/m) \max_{l=K^*+1,\dots,k} |W^*(l-T^*)|
$$

and by Proposition 1.3.2 we have $\big|W(T^*)-W(K^*)\big| = \mathcal{O}_P\big(\sqrt{\log(T^*)}\big) = \mathcal{O}_P\big(\sqrt{H-K^*}\big)$. Thus, by (1.2.3), $H \sim m$ and $(H-K^*)/m \to 0$ it holds that

$$
\max_{k=K^*+1,\dots,H} \frac{|T_k - T_{K^*} - \Lambda_k|}{h(G(k/m))}
$$

$$
= \max_{k=K^*+1,\dots,H} \frac{g(k/m)\, H^\kappa}{\sqrt{m}\, h(G(k/m))}\, \mathcal{O}_P(1)
$$

$$
+ \max_{k=K^*+1,\dots,H} \frac{g(k/m)\, \sqrt{H-K^*}}{\sqrt{m}\, h(G(k/m))}\, \mathcal{O}_P(1)
$$

$$
= o_P(1),
$$

which completes the proof.   □

We are now in the position to prove the main theorem of this section.

Proof of Theorem 1.5.1. As done before, we show the assertion for the positive stopping time τ_m^+ (i.e. on $\delta > 0$) in detail and the proof for the negative stopping time τ_m^- (i.e. for $\delta < 0$) follows analogously. The first step of the proof is to show that

$$
P\big(\tau_m^+ \leq H\big) = P\Big(\max_{k=m,\dots,H} T_k/h(G(k/m)) > 1 \Big) \to P\big(W(1) > -x\big).
$$

Once again we decompose the test statistic into its stochastic and its deterministic parts, where, heuristically speaking, the former shall give us $W(1)$ and the latter shall give us $-x$. By Lemma 1.5.6 and an approximation of the maximum in an obvious manner we have

$$\frac{T_{K^*}}{h(G(H/m))} + \frac{\Lambda_H}{h(G(H/m))} + o_P(1)$$

$$\leq \max_{k=m,\dots,H} \frac{T_k}{h(G(k/m))} \qquad (1.5.13)$$

$$\leq \max_{k=m,\dots,H} \frac{T_{K^*}}{h(G(k/m))} + \max_{k=K^*+1,\dots,H} \frac{\Lambda_k}{h(G(k/m))} + o_P(1),$$

where in the last line, we also used the fact that $\Lambda_k = 0$ for $k \leq K^*$, but $\Lambda_k > 0$ for $k > K^*$. The continuity of $h(G(t))$ at $t = 1$ in combination with Lemma 1.5.5 yields

$$\max_{k=m,\dots,H} \frac{T_k}{h(G(k/m))} = \frac{T_{K^*}}{h(G(K^*/m))} + 1 + \frac{x\sqrt{G(1)}}{h(G(1))} + o_P(1).$$

Now, along the lines of the proof of Theorem 1.3.1 we obtain

$$\frac{T_{K^*}}{h(G(K^*/m))} = \frac{\int_0^{K^*/m} g(x)\,dW_m(x)}{h(G(K^*/m))} + o_P(1)$$

and by $K^* \sim m$ it holds that

$$\frac{\int_0^{K^*/m} g(x)\,dW_m(x)}{h(G(K^*/m))} \overset{D}{=} \frac{\sqrt{G(K^*/m)}\,W(1)}{h(G(K^*/m))} = \frac{\sqrt{G(1)}\,W(1)}{h(G(1))} + o_P(1).$$

Hence, under the alternative H_1^+ it holds that

$$P\big(\tau_m^+ \leq H\big) \rightarrow P(W(1) > -x) = \phi(x). \qquad (1.5.14)$$

Similarly, under the alternative H_1^- we have

$$\frac{-T_{K^*}}{h(G(H/m))} + \frac{-\Lambda_H}{h(G(H/m))} + o_P(1)$$

$$\leq \max_{k=m,\dots,H} \frac{-T_k}{h(G(k/m))}$$

$$\leq \max_{k=m,\dots,H} \frac{-T_{K^*}}{h(G(k/m))} + \max_{k=K^*+1,\dots,H} \frac{-\Lambda_k}{h(G(k/m))} + o_P(1),$$

hence by Theorem 1.3.1 and Lemma 1.5.5 we obtain

$$\max_{k=m,\dots,H} \frac{-T_k}{h(G(k/m))}$$

$$= \frac{-\int_0^{K^*/m} g(x)\,dW_m(x)}{h(G(K^*/m))} + 1 + \frac{\sqrt{G(1)}}{h(G(1))}\,x + o_P(1), \qquad (1.5.15)$$

and therefore

$$P\big(\tau_m^- \leq H\big) \rightarrow P(-W(1) > -x) = \phi(x). \qquad (1.5.16)$$

Finally, by the fact that $\nu\, m^{(1/2+\gamma)/(1+\gamma)}$ is non-negative we see that

$$P\big(\tau_m^+ \le H\big) = P\Big(\tau_m^+ - T^* \le \big[\nu\, m^{\frac{1/2+\gamma}{1+\gamma}}\big]\Big)$$

$$= P\Big(\big(\tau_m^+ - T^*\big)_+ \le \nu\, m^{\frac{1/2+\gamma}{1+\gamma}} + \mathcal{O}(1)\Big)$$

$$= P\Bigg(\frac{(\tau_m^+ - T^*)_+^{1+\gamma}}{m^{1/2+\gamma}} \le \nu^{1+\gamma} + o(1)\Bigg)$$

$$= P\Bigg(\frac{(\tau_m^+ - T^*)_+^{1+\gamma}\, g(1)\, |\delta|}{m^{1/2+\gamma}\,(1+\gamma)\, b\,\sqrt{G(1)}} - \frac{h(G(1))}{\sqrt{G(1)}} \le x + o(1)\Bigg).$$

Combining this with the convergence obtained in (1.5.14) shows the assertion for the one-sided, positive stopping time τ_m^+. Replacing τ_m^+ by τ_m^- in the calculation above and combining this with (1.5.16) completes the proof.   □

Finally, we give the proof of Corollary 1.5.3.

Proof of Corollary 1.5.3. Note that for $i = 1, 2$, $\chi_i \ge -h(G(1))/\sqrt{G(1)}$, hence the following applications of Theorem 1.5.1 hold true: For the one-sided, positive stopping time τ_m^+ we have under H_1^+ that

$$P\big(\tau_m^+ - q(\chi_1) \le T^* < \tau_m^+\big)$$

$$= P\big(0 < \tau_m^+ - T^* \le q(\chi_1)\big)$$

$$= P\big(\tau_m^+ - T^* \le q(\chi_1)\big) - P\big(\tau_m^+ - T^* \le 0\big)$$

$$= P\Big(\big(\tau_m^+ - T^*\big)_+^{1+\gamma} \le q(\chi_1)^{1+\gamma}\Big) - P\Big(\big(\tau_m^+ - T^*\big)_+^{1+\gamma} \le 0\Big)$$

$$\to 1 - \tilde{\alpha} + \phi\big(-h(G(1))/\sqrt{G(1)}\big) - \phi\big(-h(G(1))/\sqrt{G(1)}\big)$$

$$= 1 - \tilde{\alpha},$$

where the same holds true under the one-sided alternative H_1^- for the stopping time τ_m^-. We demonstrate the corresponding assertion for the two-sided stopping time $\tilde{\tau}_m$ for the case of $\delta > 0$. We decompose the suggested confidence interval as follows:

$$P\big(\tilde{\tau}_m - q(\chi_2) \le T^* < \tilde{\tau}_m\big)$$

$$= P\big(T^* < \tilde{\tau}_m \le T^* + q(\chi_2)\big)$$

$$= P\big(\min\{\tau_m^+, \tau_m^-\} \le T^* + q(\chi_2)\big) - P\big(\min\{\tau_m^+, \tau_m^-\} \le T^*\big)$$

$$= P\big(\tau_m^+ \le T^* + q(\chi_2)\big) + P\big(\tau_m^- \le T^* + q(\chi_2)\big) \qquad (1.5.17)$$

$$\quad - P\big(\max\{\tau_m^+, \tau_m^-\} \le T^* + q(\chi_2)\big)$$

$$\quad - \Big(P\big(\tau_m^+ \le T^*\big) + P\big(\tau_m^- \le T^*\big) - P\big(\max\{\tau_m^+, \tau_m^-\} \le T^*\big)\Big)$$

$$\ge P\big(T^* < \tau_m^+ \le T^* + q(\chi_2)\big) - P\big(\tau_m^- \le T^*\big),$$

where for the first expression, since $\delta > 0$, we can make use of Theorem 1.5.1 once more. Similar as above we have

$$
\begin{aligned}
P\big(T^* < \tau_m^+ \le T^* + q(\chi_2)\big) & \\
\to\ & 1 - \tilde{\alpha} + 2\phi(-h(G(1))/\sqrt{G(1)}) - \phi(-h(G(1))/\sqrt{G(1)}) \\
=\ & 1 - \tilde{\alpha} + \phi(-h(G(1))/\sqrt{G(1)}).
\end{aligned}
$$

On the other hand, by the early-change assumption (see (1.5.15) for details) we have

$$
\max_{k=m,\dots,T^*} \frac{-T_k}{h(G(k/m))} = \frac{-\int_0^{K^*/m} g(x)\, dW_m(x)}{h(G(K^*/m))} + o_P(1).
$$

(Note that by $k < T^*$ the deterministic term in (1.5.15) disappears.) Further, by

$$
\frac{-\int_0^1 g(x)\, dW_m(x)}{h(G(1))} \overset{D}{=} \frac{-W(1)\sqrt{G(1)}}{h(G(1))}
$$

it holds that

$$
\begin{aligned}
P\big(\tau_m^- \le T^*\big) &\to P\big(-W(1) > h(G(1))/\sqrt{G(1)}\big) \\
&= \phi\big(-h(G(1))/\sqrt{G(1)}\big).
\end{aligned}
$$

Combining the two relations above yields the confidence interval under H_1 in case of positive δ. For negative δ, the assertion follows in the same way, yet by exchanging the roles of the stopping times τ_m^- and τ_m^+. $\qquad\square$

1.5.2 Asymptotic normality of the delay times for unknown in-control parameters

In this section we prove a theorem corresponding to Theorem 1.5.1 for the case of unknown in-control parameters a and b.

Theorem 1.5.7. *With the notation and assumptions of Section 1.1 and Section 1.2 as well as $h(\tilde{G}(t))$ being continuous at $t = 1$, (1.5.1) holding true with δ being constant,*

$$\sup_{1 \leq \xi \leq \tilde{H}/m} |g'(\xi)| = \mathcal{O}(1) \tag{1.5.18}$$

(where $\tilde{H}$ is defined in (1.5.21) below) and the early-change assumption

$$T^* = m + o(m) \tag{1.5.19}$$

it holds for all $x \geq -h(\tilde{G}(1))/\sqrt{\tilde{G}(1)}$ under H_1^+ that

$$\lim_{m \to \infty} P\left(\frac{(\hat{\tau}_m^+ - T^*)_+^{1+\gamma} \left(g(1) - \int_0^1 g(x)\,dx\right)\delta}{m^{1/2+\gamma}\,(1+\gamma)\,b\,\sqrt{\tilde{G}(1)}} - \frac{h(\tilde{G}(1))}{\sqrt{\tilde{G}(1)}} \leq x \right) = \phi(x)$$

and under H_1^- that

$$\lim_{m \to \infty} P\left(\frac{(\hat{\tau}_m^- - T^*)_+^{1+\gamma} \left(g(1) - \int_0^1 g(x)\,dx\right)|\delta|}{m^{1/2+\gamma}\,(1+\gamma)\,b\,\sqrt{\tilde{G}(1)}} - \frac{h(\tilde{G}(1))}{\sqrt{\tilde{G}(1)}} \leq x \right) = \phi(x).$$

By Slutsky's lemma the convergences above still hold true if we replace b and δ by consistent estimates $\hat{b}$ and $\hat{\delta}$, where possible choices are suggested in Remark 1.2.6 and Lemma 1.5.8. (See also the discussion on $\hat{\delta}$ of Section 1.5.1.)

Lemma 1.5.8. *With the notation and assumptions of Section 1.1 let*

$$\hat{\delta}_{m,k} = \frac{\sum_{i=1}^k ((i-m)/m)_+^\gamma \, (Z_i - Z(m)/m)}{\sum_{i=1}^k ((i-m)/m)_+^{2\gamma}}.$$

If in addition for some $\zeta > 1$ it holds that

$$T^* - m = o(m^\zeta),$$

then we have under either of the alternatives H_1^+, H_1^- or H_1 that

$$\hat{\delta}_{m,m+m^\zeta} = \delta + o_P(1).$$

The proof of Lemma 1.5.8 is given at the very end of this section.

Combining Theorem 1.5.1 and Lemma 1.5.8 gives us the following asymptotic confidence intervals for the change point:

Corollary 1.5.9. *Let the assumptions of Theorem 1.5.1 hold true. On setting*

$$q(\chi_i) = \left(\left(\frac{h(\tilde{G}(1))}{\sqrt{\tilde{G}(1)}} + \chi_i \right) \frac{(1+\gamma)\,\hat{b}_m\,\sqrt{\tilde{G}(1)}\,m^{1/2+\gamma}}{g(1)\,|\hat{\delta}|} \right)^{\frac{1}{1+\gamma}},$$

where $\hat{\delta} \xrightarrow{P} \delta$ *and* $\hat{b}_m \xrightarrow{P} b$ *are consistent estimates (e.g.* $\hat{\delta}_{m,m+m\varsigma}$ *as in Lemma 1.5.8 with* $\varsigma > 1$ *and* $\hat{b}_m$ *as in Remark 1.2.6) and*

$$\chi_i = \begin{cases} \phi^{-1}\big(1 - \tilde{\alpha} + \phi(-h(\tilde{G}(1))/\sqrt{\tilde{G}(1)})\big) & \text{if } i = 1, \\ \phi^{-1}\big(1 - \tilde{\alpha} + 2\phi(-h(\tilde{G}(1))/\sqrt{\tilde{G}(1)})\big) & \text{if } i = 2, \end{cases}$$

with

$$\tilde{\alpha} \in \begin{cases} \big(\phi(-h(\tilde{G}(1))/\sqrt{\tilde{G}(1)}),\, 1\big) & \text{for } i = 1, \\ \big(2\,\phi(-h(\tilde{G}(1))/\sqrt{\tilde{G}(1)}),\, 1\big) & \text{for } i = 2, \end{cases} \tag{1.5.20}$$

we obtain for $\hat{\tau}_m^+$ *and* $\hat{\tau}_m^-$ *as in Section 1.2.1 and* $\check{\tau}_m := \min\{\hat{\tau}_m^+, \hat{\tau}_m^-\}$ *that*

$$P\big(\hat{\tau}_m^+ - q(\chi_1) \leq T^* < \hat{\tau}_m^+\big) = 1 - \tilde{\alpha} + o(1),$$
$$P\big(\hat{\tau}_m^- - q(\chi_1) \leq T^* < \hat{\tau}_m^-\big) = 1 - \tilde{\alpha} + o(1),$$
$$P\big(\check{\tau}_m - q(\chi_2) \leq T^* < \check{\tau}_m\big) \geq 1 - \tilde{\alpha} + o(1)$$

under H_1^+, H_1^- *or* H_1, *respectively.*

The proof of Corollary 1.5.9 follows along the lines of the proof of Corollary 1.5.3 and is therefore omitted. Hence, we turn to the proof of Theorem 1.5.7. As in the previous section the proof of the main theorem is subdivided into three steps and to shorten the upcoming proofs we introduce the following notation: For $x \geq -h(\tilde{G}(1))/\sqrt{\tilde{G}(1)}$ and $K^* = [T^*]$ let

$$\tilde{H} := K^* + \left[\tilde{\nu}\,m^{\frac{1/2+\gamma}{1+\gamma}}\right], \tag{1.5.21}$$

where

$$\tilde{\nu} := \left(\left(1 + x\,\frac{\sqrt{\tilde{G}(1)}}{h(\tilde{G}(1))}\right) \frac{(1+\gamma)\,b\,h(\tilde{G}(1))}{(g(1) - \int_0^1 g(x)dx)\,|\delta|} \right)^{\frac{1}{1+\gamma}}.$$

Further, let

$$\hat{\Lambda}_k := \frac{\sum_{i=K^*+1}^k \delta((i - T^*)/m)^\gamma \big(g(i/m) - \sum_{j=1}^k g(j/m)/k\big)}{\sqrt{m}\,b} \tag{1.5.22}$$

denote the deterministic perturbation of the detectors.

Lemma 1.5.10. *Under the assumptions of Theorem 1.5.7 and with* K^*, $\tilde{H}$ *and* $\hat{\Lambda}_k$ *as in* (1.5.6), (1.5.21) *and* (1.5.22) *it holds for* $x \geq -h(\tilde{G}(1))/\sqrt{\tilde{G}(1)}$, *as* $m \to \infty$:

$$\frac{\tilde{H} - T^*}{m} \to 0, \tag{1.5.23}$$

$$\tilde{H} \sim K^*, \tag{1.5.24}$$

$$\frac{\operatorname{sgn}(\delta)\, \hat{\Lambda}_{\tilde{H}}}{h(\tilde{G}(\tilde{H}/m))} \to 1 + x\, \frac{\sqrt{\tilde{G}(1)}}{h(\tilde{G}(1))}, \tag{1.5.25}$$

$$\max_{k=K^*+1,\ldots,\tilde{H}} \frac{\operatorname{sgn}(\delta)\, \hat{\Lambda}_k}{h(\tilde{G}(k/m))} \to 1 + x\, \frac{\sqrt{\tilde{G}(1)}}{h(\tilde{G}(1))}. \tag{1.5.26}$$

Proof of Lemma 1.5.10. The first two assertions are immediate consequences of the definition of $\tilde{H}$ and of the early-change assumption (1.5.19). On proving (1.5.25) we decompose $\hat{\Lambda}_{\tilde{H}}$ in the typical manner, i.e.

$$\hat{\Lambda}_{\tilde{H}} = \Lambda_{\tilde{H}} - \frac{\sum_{i=K^*+1}^{\tilde{H}} \delta((i - T^*)/m)^\gamma \sum_{j=1}^{\tilde{H}} g(j/m)/\tilde{H}}{\sqrt{m}\, b}, \tag{1.5.27}$$

where Λ_k is defined in (1.5.8). Along the lines of the proof of Lemma 1.5.5 one can show that the $\Lambda_{\tilde{H}}$ converges towards $\delta \tilde{\nu}^{1+\gamma} g(1)/(b(1+\gamma))$. As to the second term we note that, for one thing, by (1.5.24)

$$\sum_{j=1}^{\tilde{H}} \frac{g(j/m)}{\tilde{H}} = \frac{m}{\tilde{H}} \left(\int_0^{\tilde{H}/m} g(x)\, dx + \mathcal{O}\big(g(\tilde{H}/m)/m\big) \right) = \int_0^1 g(x)\, dx + o(1) \tag{1.5.28}$$

and, for another thing, similar as in Lemma 1.5.5

$$\frac{\sum_{i=K^*+1}^{\tilde{H}} \delta((i - T^*)/m)^\gamma}{\sqrt{m}\, b\, h(\tilde{G}(\tilde{H}/m))} = \frac{\delta\, \tilde{\nu}^{1+\gamma}}{b\,(1+\gamma)\, h(\tilde{G}(1))} + o(1).$$

Combining the two assertions above yields

$$\frac{\operatorname{sgn}(\delta)\, \hat{\Lambda}_{\tilde{H}}}{h(\tilde{G}(\tilde{H}/m))} \to \frac{\operatorname{sgn}(\delta)\, \delta\, \tilde{\nu}^{1+\gamma} \big(g(1) - \int_0^1 g(x)\, dx\big)}{(1+\gamma)\, b\, h(\tilde{G}(1))} = 1 + x\, \frac{\sqrt{\tilde{G}(1)}}{h(\tilde{G}(1))}.$$

By elementary calculations we see that $\operatorname{sgn}(\delta)\, \hat{\Lambda}_k - \operatorname{sgn}(\delta)\, \hat{\Lambda}_{k-1} > 0$, hence $\operatorname{sgn}(\delta)\, \hat{\Lambda}_k$ is increasing and the convergence of (1.5.26) follows along the lines of (1.5.12).   □

Lemma 1.5.11. *Under the assumptions of Theorem 1.5.7 it holds that, as* $m \to \infty$,

$$\max_{k=m,\ldots,\tilde{H}} \frac{|\tilde{T}_k - \tilde{T}_{K^*} - \hat{\Lambda}_k|}{h(\tilde{G}(k/m))} = o_P(1),$$

where $\tilde{T}_k := \hat{T}_k\, \hat{b}_k/b$ *shall denote the test statistic with an estimated location but a non-estimated scale parameter.*

Proof of Lemma 1.5.11. We demonstrate the proof for $k = K^* + 1, \ldots, \tilde{H}$; for $k = m, \ldots, K^*$ the assertion follows by the same arguments, but in a more simple manner. In order to make use of the results for known in-control parameters, we detach the proportion of the detectors which is due to the estimation: For $k = K^* + 1, \ldots, \tilde{H}$ we have

$$
\tilde{T}_k - \tilde{T}_{K^*} - \hat{\Lambda}_k
$$
$$
= T_k - T_{K^*} - \Lambda_k
$$
$$
- \frac{\sum_{i=1}^{k} g(i/m)\big(\hat{a}_k - a - \delta \sum_{j=K^*+1}^{k}((j - T^*)/m)^\gamma /k\big)}{\sqrt{m}\,b}
$$
$$
+ \frac{\sum_{i=1}^{K^*} g(i/m)\big(\hat{a}_{K^*} - a\big)}{\sqrt{m}\,b}
$$
$$
= T_k - T_{K^*} - \Lambda_k
$$
$$
- \frac{\sum_{i=1}^{k} g(i/m)\,(b\,Y(T^*) + b^*\,Y^*(k - T^*))}{k\,\sqrt{m}\,b}
$$
$$
+ \frac{\sum_{i=1}^{K^*} g(i/m)\,Y(K^*)}{K^*\,\sqrt{m}},
$$

where as in Lemma 1.5.6 (taking Assumption 1.2.4 into account) we have

$$
\max_{k=m,\ldots,\tilde{H}} \frac{|T_k - T_{K^*} - \Lambda_k|}{h(\tilde{G}(k/m))} = o_P(1).
$$

We look at the additional terms, which arise from the estimation of a: By the invariance principles (1.1.2) and (1.1.3) an approximation of the respective sums in the manner of (1.5.28) we have uniformly for $k = K^* + 1, \ldots, \tilde{H}$

$$
\frac{\sum_{i=1}^{K^*} g(i/m)\,Y(K^*)}{K^*\,\sqrt{m}} - \frac{\sum_{i=1}^{k} g(i/m)\,Y(K^*)}{k\,\sqrt{m}} = o_P(1)
$$

and by Proposition 1.3.2 we have uniformly for $k = K^* + 1, \ldots, \tilde{H}$

$$
\frac{\sum_{i=1}^{k} g(i/m)\,Y(K^*)}{k\,\sqrt{m}} - \frac{\sum_{i=1}^{k} g(i/m)\,Y(T^*)}{k\,\sqrt{m}} = o_P(1).
$$

Further, taking (1.1.3) into account, it holds uniformly for $k = K^* + 1, \ldots, \tilde{H}$ that

$$
\frac{\sum_{i=1}^{k} g(i/m)\,b^*\,Y^*(k - T^*)}{k\,\sqrt{m}\,b} = \mathcal{O}_P\big((\tilde{H}^\kappa + \sqrt{\tilde{H} - K^*})/\sqrt{m}\big) = o_P(1),
$$

which completes the proof. $\qquad\qquad\qquad\qquad\qquad\qquad\qquad\qquad\square$

By the means of Lemma 1.5.10 and Lemma 1.5.11, the proof of Theorem 1.5.7 follows along the lines of the proof of Theorem 1.5.1, hence it is omitted. Finally, we give the proof of the consistency of the estimate $\hat{\delta}$.

Proof of Lemma 1.5.8. Note that by $T^* - m = o(m^\varsigma)$ it holds for sufficiently large m, that $T^* < m + m^\varsigma$. We start by separating the stochastic and the deterministic parts of $\hat{\delta}_{m,m+m^\varsigma}$:

$$
\begin{aligned}
\sum_{i=1}^{m+m^\varsigma} & ((i-m)/m)_+^\gamma \left(Z(i) - Z(i-1) - Z(m)/m \right) \\
= & \sum_{i=m+1}^{K^*} ((i-m)/m)^\gamma \, b \left(Y(i) - Y(i-1) - Y(m)/m \right) \\
& + ((K^*+1-m)/m)^\gamma \, b \left(Y(T^*) - Y(K^*) - Y(m)/m \right) \\
& + ((K^*+1-m)/m)^\gamma \, b^* \, Y(K^*+1-T^*) \\
& + \sum_{i=K^*+2}^{m+m^\varsigma} ((i-m)/m)^\gamma \left(b^* \, Y^*(i-T^*) - b^* Y^*(i-1-T^*) \right) \\
& - \sum_{i=K^*+2}^{m+m^\varsigma} ((i-m)/m)^\gamma \, b \, Y(m)/m \\
& + \delta \sum_{i=K^*+1}^{m+m^\varsigma} ((i-m)/m)^\gamma ((i-T^*)/m)^\gamma,
\end{aligned}
$$
(1.5.29)

where that the last (deterministic) sum converges towards δ and the remaining (stochastic) terms do not contribute to the asymptotic. For the latter, we confine ourselves to showing that the fourth sum is of order $o_P(1)$; the same follows for the remaining four stochastic terms in a similar manner. As usual, the first step is to make use of the invariance principle (1.1.3) to replace $Y^*(i-T^*)$ by a Wiener process. Note that

$$
\sum_{i=1}^{m+m^\varsigma} ((i-m)/m)_+^\gamma \simeq m^{\varsigma(\gamma+1)-\gamma} = m^{\gamma(\varsigma-1)+\varsigma},
$$

$$
\sum_{i=1}^{m+m^\varsigma} ((i-m)/m)_+^{2\gamma} \simeq m^{\varsigma(2\gamma+1)-2\gamma} = m^{2\gamma(\varsigma-1)+\varsigma},
$$

which gives us via Remark 1.3.3 the following rate of convergence:

$$
\begin{aligned}
& \frac{\sum_{i=K^*+2}^{m+m^\varsigma} ((i-m)/m)_+^\gamma (Y^*(i-T^*) - Y^*(i-1-T^*))}{\sum_{i=1}^{m+m^\varsigma} ((i-m)/m)_+^{2\gamma}} \\
= & \frac{\sum_{i=K^*+2}^{m+m^\varsigma} ((i-m)/m)_+^\gamma (W^*(i-T^*) - W^*(i-1-T^*))}{\sum_{i=1}^{m+m^\varsigma} ((i-m)/m)_+^{2\gamma}} \\
& + \frac{m^{\gamma(\varsigma-1)} \, m^{\varsigma \kappa}}{m^{2\gamma(\varsigma-1)+\varsigma}} \, \mathcal{O}_P(1) \\
= & \frac{m^{(\gamma(\varsigma-1)+\varsigma)} \, \sqrt{\log(m)}}{m^{2\gamma(\varsigma-1)+\varsigma}} \, \mathcal{O}_P(1) + o_P(1) \\
= & \, o_P(1),
\end{aligned}
$$

where in the last two equalities we used $\zeta > 1$. Now, we show that the deterministic part of (1.5.29) (i.e. the last summand) converges to δ, which is equivalent to showing that

$$\frac{\sum_{i=1}^{m+m^\zeta} (i-m)_+^\gamma \big((i-m)_+^\gamma - (i-T^*)_+^\gamma\big)}{\sum_{i=1}^{m+m^\zeta} (i-m)_+^{2\gamma}} \to 0.$$

The latter term vanishes for $i \leq K^* = [T^*]$, hence we have

$$\sum_{i=1}^{m+m^\zeta} (i-m)_+^\gamma \big((i-m)_+^\gamma - (i-T^*)_+^\gamma\big)$$

$$= \sum_{i=m+1}^{K^*} (i-m)_+^{2\gamma} + \sum_{i=K^*+1}^{m+m^\zeta} (i-m)_+^\gamma \big((i-m)_+^\gamma - (i-T^*)_+^\gamma\big).$$

On the one hand, we have

$$\frac{\sum_{i=m+1}^{K^*} (i-m)_+^{2\gamma}}{\sum_{i=1}^{m+m^\zeta} (i-m)_+^{2\gamma}} = \mathcal{O}\Big(\big((T^*-m)/m^\zeta\big)^{2\gamma+1}\Big) = o(1).$$

On the other hand, we have by the Cauchy–Schwarz inequality and the mean value theorem for some $\xi_i \in [i-T^*, i-m]$ that

$$\frac{\sum_{i=K^*+2}^{m+m^\zeta} (i-m)_+^\gamma \big((i-m)_+^\gamma - (i-T^*)_+^\gamma\big)}{\sum_{i=1}^{m+m^\zeta} (i-m)_+^{2\gamma}}$$

$$\leq \frac{\Big(\sum_{i=K^*+2}^{m+m^\zeta} (i-m)_+^{2\gamma}\Big)^{1/2} \Big(\sum_{i=K^*+2}^{m+m^\zeta} \big((i-m)_+^\gamma - (i-T^*)_+^\gamma\big)^2\Big)^{1/2}}{\sum_{i=1}^{m+m^\zeta} (i-m)_+^{2\gamma}}$$

$$= \mathcal{O}\big(m^{\zeta(2\gamma+1)/2}/m^{\zeta(2\gamma+1)}\big) \Big(\sum_{i=1}^{m+m^\zeta} \big(\xi_i^{\gamma-1}(T^*-m)\big)^2\Big)^{1/2}$$

$$= \mathcal{O}\big(m^{-\zeta(2\gamma+1)/2}\big) \mathcal{O}\big(m^{\zeta(2\gamma-1)/2}(T^*-m)\big)$$

$$= \mathcal{O}\big((T^*-m)/m^\zeta\big)$$

$$= o(1),$$

where the last equality follows by our assumption on ζ. $\qquad\square$

1.6 Finite sample behavior

In this section we present some simulation results in order to demonstrate the finite sample behavior of the suggested procedures. As introduced in Example 1.1.1 we consider a process $Z_i = \varepsilon_i + \delta((i - T^*)/m)_+^\gamma$, where the innovations ε_i are independently, identically distributed. For the results below we choose $\varepsilon_i \sim \text{Exp}(1)$. We consider different lengths of the training period m, where in each case the process is monitored for time period of $10\,m$ and the null hypothesis is kept if no change has been indicated until then. For the results under the alternative we choose $T^* = m + m^{0.5}$, which fulfills the early-change assumption of Theorem 1.5.1 and Theorem 1.5.7. Moreover, we choose different sizes for the parameters δ and γ, varying in between 0.5 and 3 or 0.2 and 2, respectively.

For the testing procedure we follow Remark 1.2.2, Remark 1.2.3 and Remark 1.2.5: As a weight function we use $g(t) = t^\lambda$, where we consider $\lambda = 0.2$, $\lambda = 0.5$ and $\lambda = 1.5$, and as a threshold function we use $h(t) = c\,t$ where c is adjusted such that we attain the prescribed level α asymptotically. In case of unknown in-control parameters, the parameter b is estimated (non-sequentially) by

$$\hat{b}_k = \hat{b}_m = \frac{1}{\hat{m}(m - \hat{m} + 1)} \sum_{j=\hat{m}}^{m} \big(Z(j) - Z(j - \hat{m}) - \hat{m}\,Z(m)/m\big)^2,$$

where $\hat{m} = m^{0.25}$ (see Remark 1.2.6). Each result is based on 5000 repetitions.

Table 1.1 and Table 1.2 show the relative frequency of a false alarm for known or unknown in-control parameters, respectively, with a prescribed asymptotic level of $\alpha = 5\%$ and $\alpha = 10\%$. We see that the asymptotic level is (mostly) well attained, where, as one would expect, in case of known in-control parameters the simulated sizes are slightly closer to the prescribed asymptotic size.

For the simulation results under the alternative we restrict ourselves to the case of $\alpha = 5\%$. The consistency of the testing procedure is investigated in Table 1.3 and Table 1.4 (for known in-control parameters) and Table 1.5, Table 1.6 and Table 1.7 (for unknown in-control parameters), where each table gives the power of the procedure for a fixed λ. In case of known in-control parameters and $\lambda = 0.2$ all changes were detected, hence the corresponding table is omitted. Note that the case of $\gamma = 0.5$ and $\lambda = 1.5$ does not fulfill the assumptions of the consistency results obtained in Theorem 1.4.1 and Theorem 1.4.3 (i.e. $\lambda \leq \gamma$, see Remark 1.4.2, Remark 1.4.5 and Section 1.7) which explains the poorer detection rate in this case.

Finally, we illustrate the asymptotic normality of the delay times (see Theorem 1.5.1 and Theorem 1.5.7) by comparing histograms of the standardized delay times with the density function of the standard normal distribution. For training periods of length $m = 500$ and $m = 1000$ we consider the parameter combinations $\lambda = 0.2$, $\gamma = 0.5$ and $\lambda = \gamma = 0.5$. For larger λ and γ the results become worse and the histograms have a stronger tendency to the right, which indicates that a smaller λ might be the better choice in practice. In Figure 1.2, the parameter δ was estimated by $\hat{\delta}_{m,m+m\zeta}$ of Lemma 1.5.8, where we choose $\zeta = 1.05$, whereas for Figure 1.1 we used the true values of δ.

m	$\alpha = 5\%$			$\alpha = 10\%$		
	$\lambda = 0.2$	$\lambda = 0.5$	$\lambda = 1.5$	$\lambda = 0.2$	$\lambda = 0.5$	$\lambda = 1.5$
25	.0576	.0590	.0684	.1036	.0956	.1044
50	.0604	.0644	.0572	.1110	.1006	.1054
100	.0574	.0592	.0524	.0952	.1076	.1038
250	.0500	.0478	.0606	.1030	.0956	.1048
500	.0538	.0534	.0558	.1008	.0986	.1020
1000	.0512	.0628	.0610	.0968	.1044	.0990

Table 1.1: Relative frequency of a false alarm under the null hypothesis for known in-control parameters

m	$\alpha = 5\%$			$\alpha = 10\%$		
	$\lambda = 0.2$	$\lambda = 0.5$	$\lambda = 1.5$	$\lambda = 0.2$	$\lambda = 0.5$	$\lambda = 1.5$
25	.0864	.0922	.0730	.1584	.1624	.1212
50	.0564	.0650	.0518	.1112	.1268	.0916
100	.0620	.0616	.0576	.1092	.1138	.0928
250	.0520	.0450	.0392	.0932	.1016	.0846
500	.0454	.0480	.0340	.1054	.1060	.0928
1000	.0518	.0452	.0458	.0982	.1022	.0816

Table 1.2: Relative frequency of a false alarm under the null hypothesis for unknown in-control parameters for $\hat{m} = m^{0.25}$

$\lambda = 0.5$	$\gamma = 0.5$			$\gamma = 2$		
m	$\delta = 0.5$	$\delta = 1$	$\delta = 2$	$\delta = 0.5$	$\delta = 1$	$\delta = 2$
25	.0872	1	1	1	1	1
50	.9950	1	1	1	1	1
100	1	1	1	1	1	1
250	1	1	1	1	1	1
500	1	1	1	1	1	1
1000	1	1	1	1	1	1

Table 1.3: Relative frequency of a correct detection of a change for known in-control parameters for $\lambda = 0.5$, $T^* = m + m^{0.5}$ and $\alpha = 5\%$

$\lambda = 1.5$	$\gamma = 0.5$			$\gamma = 2$		
m	$\delta = 0.5$	$\delta = 1$	$\delta = 2$	$\delta = 0.5$	$\delta = 1$	$\delta = 2$
25	.0690	.0680	.1184	.0610	.1316	1
50	.0656	.0750	.9792	.0572	1	1
100	.0620	.2076	1	.9660	1	1
250	.1040	1	1	1	1	1
500	.6248	1	1	1	1	1
1000	.1	1	1	1	1	1

Table 1.4: Relative frequency of a correct detection of a change for known in-control parameters for $\lambda = 1.5$, $T^* = m + m^{0.5}$ and $\alpha = 5\%$

$\lambda = 0.2$	$\gamma = 0.5$			$\gamma = 2$		
m	$\delta = 0.5$	$\delta = 1$	$\delta = 2$	$\delta = 0.5$	$\delta = 1$	$\delta = 2$
25	.9170	1	1	1	1	1
50	.9978	1	1	1	1	1
100	1	1	1	1	1	1
250	1	1	1	1	1	1
500	1	1	1	1	1	1
1000	1	1	1	1	1	1

Table 1.5: Relative frequency of a correct detection of a change for unknown in-control parameters for $\lambda = 0.2$, $T^* = m + m^{0.5}$, $\hat{m} = m^{0.25}$ and $\alpha = 5\%$

$\lambda = 0.5$	$\gamma = 0.5$			$\gamma = 2$		
m	$\delta = 0.5$	$\delta = 1$	$\delta = 2$	$\delta = 0.5$	$\delta = 1$	$\delta = 2$
25	.2824	.9984	1	1	1	1
50	.9256	1	1	1	1	1
100	1	1	1	1	1	1
250	1	1	1	1	1	1
500	1	1	1	1	1	1
1000	1	1	1	1	1	1

Table 1.6: Relative frequency of a correct detection of a change for unknown in-control parameters for $\lambda = 0.5$, $T^* = m + m^{0.5}$, $\hat{m} = m^{0.25}$ and $\alpha = 5\%$

$\lambda = 1.5$	$\gamma = 0.5$			$\gamma = 2$		
m	$\delta = 0.5$	$\delta = 1$	$\delta = 2$	$\delta = 0.5$	$\delta = 1$	$\delta = 2$
25	.0960	.0984	.4754	.0804	.9648	1
50	.0666	.1040	.8540	.0560	1	1
100	.0772	.3394	1	.0718	1	1
250	.0828	.9068	1	1	1	1
500	.2450	1	1	1	1	1
1000	.8982	1	1	1	1	1

Table 1.7: Relative frequency of a correct detection of a change for unknown in-control parameters for $\lambda = 1.5$, $T^* = m + m^{0.5}$, $\hat{m} = m^{0.25}$ and $\alpha = 5\%$

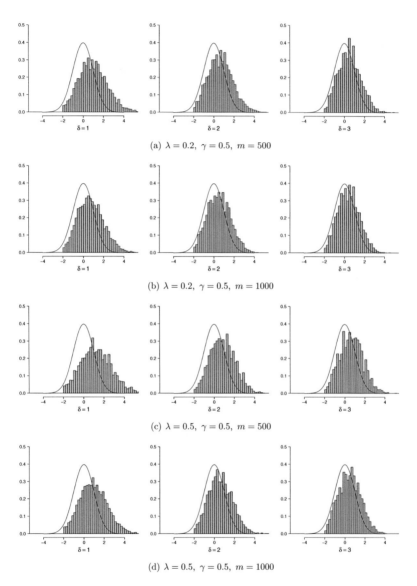

Figure 1.1: Histograms of the standardized delay times for known in-control parameters for $\alpha = 5\%$ and $T^* = m + m^{0.5}$

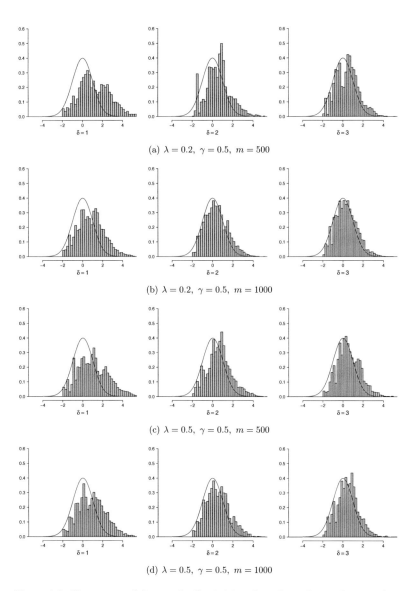

(a) $\lambda = 0.2$, $\gamma = 0.5$, $m = 500$

(b) $\lambda = 0.2$, $\gamma = 0.5$, $m = 1000$

(c) $\lambda = 0.5$, $\gamma = 0.5$, $m = 500$

(d) $\lambda = 0.5$, $\gamma = 0.5$, $m = 1000$

Figure 1.2: Histograms of the standardized delay times for unknown in-control parameters for $\alpha = 5\%$, $T^* = m + m^{0.5}$, $\hat{m} = m^{0.25}$ and $\zeta = 1.05$

1.7   Some calculations

In this section we verify the growth rates of Assumption 1.2.1 and Assumption 1.2.4 and the growth conditions of Theorem 1.4.1 and Theorem 1.4.3 for the weight and threshold functions suggested in Remark 1.2.2 and Remark 1.2.5 and the type of change suggested in Example 1.1.1 and Example 1.1.2.

Recall that $g(t) = t^\lambda$, which implies

$$G(t) = t^{1+2\lambda}/(1 + 2\lambda)$$

and

$$\tilde{G}(t) = \tilde{\lambda}\, t^{1+2\lambda}, \quad \text{where} \quad \tilde{\lambda} = \lambda^2/((1 + 2\lambda)(1 + \lambda)^2)$$

and

$$h^{(1)}(t) = c\,t,$$
$$h^{(2)}(t) = \sqrt{t}\; f^{-1}\big(\ln(t) + f(c)\big), \quad \text{where} \quad f(t) = t^2 + 2\ln(\phi(t)),$$
$$h^{(3)}(t) = \sqrt{t\,(c^2 + \ln(t))}.$$

The following remark implies $h^{(2)}(t) \sim h^{(3)}(t)$ and therefore

$$\sup_{t \geq m} t^{1/2 + \lambda} / h^{(i)}(G(t)) = \mathcal{O}(1)$$

and

$$\sup_{t \geq m} t^{1/2 + \lambda} / h^{(i)}(\tilde{G}(t)) = \mathcal{O}(1)$$

for $i = 1, 2, 3$.

Remark 1.7.1. *The function*

$$f : \mathbb{R}^{\geq 0} \to \mathbb{R}^{\geq f(0)}, \; t \mapsto t^2 + 2\ln(\phi(t)),$$

is continuous, increasing and surjective, hence it is invertible with a continuous, increasing inverse function $f^{-1} : \mathbb{R}^{\geq f(0)} \to \mathbb{R}^{\geq 0}$. We do not have an explicit expression of f^{-1}, however the following properties can easily be shown:

1. *$f(t) \sim t^2$ as $t \to \infty$, $\sup_{t \geq 1} f(t)/t^2 < \infty$,*

2. *$\sqrt{t} < f^{-1}(t) < \sqrt{t+1}$ for all $t \geq 0$,*

3. *$f^{-1}(t) \sim \sqrt{t}$ as $t \to \infty$, $\sup_{t \geq 1} f^{-1}(t)/\sqrt{t} < \infty$.*

Proof of Remark 1.7.1. The first assertion is trivial. The second assertion immediately implies the third one, hence we focus on the second assertion: Note that, on the one hand, for all $t \geq 0$

$$t = f(f^{-1}(t)) = (f^{-1}(t))^2 + 2\ln(\phi(f^{-1}(t))) < (f^{-1}(t))^2,$$

hence $\sqrt{t} < f^{-1}(t)$. On the other hand, we have for all $t \geq 0$

$$t < t + 1 + 2\ln(\phi(1)) < t + 1 + 2\ln(\phi(\sqrt{t+1})) = f(\sqrt{t+1}).$$

Via the monotonicity of f^{-1} this implies $f^{-1}(t) < \sqrt{t+1}$. $\square$

Now, we verify Assumption 1.2.1 and Assumption 1.2.4. Since $\tilde{G}(t) \simeq G(t)$ it is sufficient to consider the latter one. The following convergences hold true for $i = 1, 2, 3$ and uniformly for $t \geq m$ as $m \to \infty$:

$$\frac{t^{\kappa}\,g(t/m)}{\sqrt{m}\,h^{(i)}(G(t/m))} = \frac{t^{\kappa-1/2}\,(t/m)^{\lambda+1/2}}{h^{(i)}(G(t/m))} = o(1)$$

and

$$\sup_{0 \leq \epsilon \leq 1/m} \frac{g'(t/m - \epsilon)\,\sqrt{t\,\log\log(t/m)}}{m^{3/2}\,h^{(i)}(G(t/m))} = \frac{\sqrt{\log\log(t/m)}}{t}\,\mathcal{O}(1) = o(1).$$

Concerning Assumptions (1.2.5) and (1.2.11) we have for $h^{(1)}$:

$$\left|1 - \frac{t/m}{[t]/m}\right| = \left|\frac{[t]-t}{[t]}\right| \leq \frac{1}{m} = o(1).$$

Showing the same assertion for $h^{(2)}$ is somewhat more tedious since we do not have an explicit expression for f^{-1}. Decomposing

$$\left|1 - \frac{h^{(2)}(t/m)}{h^{(2)}([t]/m)}\right| = \left|1 - \frac{\sqrt{t/m}}{\sqrt{[t]/m}}\,\frac{f^{-1}\big(\ln(t/m) + f(c)\big)}{f^{-1}\big(\ln([t]/m) + f(c)\big)}\right|$$

we have, on the one hand,

$$\left|\sqrt{t/[t]} - 1\right| = o(1)$$

and, on the other hand, for some $\xi \in [\ln([t]/m) + f(c), \ln(t/m) + f(c)]$,

$$\left|\frac{f^{-1}\big(\ln(t/m) + f(c)\big)}{f^{-1}\big(\ln([t]/m) + f(c)\big)} - 1\right|$$

$$= \left|\frac{f^{-1}\big(\ln(t/m) + f(c)\big) - f^{-1}\big(\ln([t]/m) + f(c)\big)}{f^{-1}\big(\ln([t]/m) + f(c)\big)}\right|$$

$$= \left|\frac{\big(f^{-1}(\xi)\big)'\,\ln(t/[t])}{f^{-1}\big(\ln([t]/m) + f(c)\big)}\right|$$

$$= o(1)\,\frac{1}{f'(\xi)}$$

$$= o(1),$$

which shows the growth rates of Assumption 1.2.1 and Assumption 1.2.4 for $h^{(2)}$ and (analogously) for $h^{(3)}$.

Next, we show that the assumptions needed for the consistency of the procedure hold true with the weight and threshold functions suggested above and

$$\Delta_{m,\gamma}(i - T^*) - \Delta_{m,\gamma}(i - 1 - T^*) = \delta\big((i - T^*)/m\big)_+^\gamma,$$

for some (constant) $\delta \neq 0$ and $\gamma > 0$, where by Remark 1.4.4 we know that we can neglect the estimate $\hat{b}_N$. Conditions (1.4.2) and (1.4.4) can easily be verified: For any sequence $N > m$ it holds for $i = 1, 2, 3$ that

$$\frac{g(N/m)\sqrt{N/m}}{h^{(i)}(N/m)} = \frac{(N/m)^{\lambda+1/2}}{(N/m)^{\lambda+1/2}}\, \mathcal{O}(1) = \mathcal{O}(1).$$

On showing (1.4.3) and (1.4.5), we focus on the latter one, which is the stronger condition. If $N - K^* \to \infty$ it holds that

$$
\begin{aligned}
&\frac{|\hat{\Lambda}_N|}{h(\tilde{G}(N/m))} \\
&= \frac{|\delta| \sum_{i=K^*+1}^{N} \big((i - T^*)/m\big)^\gamma \big((i/m)^\lambda - \sum_{j=1}^{N}(j/m)^\lambda/N\big)}{\sqrt{m}\, h(\tilde{G}(N/m))} \\
&= \frac{|\delta| \sum_{i=K^*+1}^{N} \big((i - T^*)/m\big)^\gamma (N/m)^\lambda}{\sqrt{m}\, h(\tilde{G}(N/m))} \\
&\quad \times \frac{\sum_{i=K^*+1}^{N} \big((i - T^*)/m\big)^\gamma \big((i/N)^\lambda - \sum_{j=1}^{N}(j/N)^\lambda/N\big)}{\sum_{i=K^*+1}^{N} \big((i - T^*)/m\big)^\gamma} \\
&= \frac{|\delta| \sum_{i=1}^{N-T^*} \big(i/(N - T^*)\big)^\gamma/(N - K^*)\, (N - K^*)^{1+\gamma} (N/m)^\lambda}{m^{1/2+\gamma}\, h(\tilde{G}(N/m))} \\
&\quad \times \left(\frac{\sum_{i=1}^{N-T^*} i^\gamma\big((i + T^*)/N\big)^\lambda}{\sum_{i=1}^{N-T^*} i^\gamma} - \sum_{j=1}^{N}(j/N)^\lambda/N\right) \\
&\simeq \frac{|\delta|\, (N - K^*)^{1+\gamma} (N/m)^\lambda}{m^{1/2+\gamma}\, h(\tilde{G}(N/m))} \\
&\quad \times \left(\frac{\sum_{i=1}^{N-T^*} i^\gamma\big((i + T^*)/N\big)^\lambda}{\sum_{i=1}^{N-T^*} i^\gamma} - \frac{1}{\lambda + 1} + o(1)\right) \\
&=: F_1 \times (F_2 + o(1)).
\end{aligned}
\tag{1.7.1}
$$

We show that F_2 does not contribute to the asymptotic for any integer sequence N with $N - T^* \to \infty$. On the one hand we have

$$F_2 \leq 1 - \frac{1}{\lambda + 1} = \frac{\lambda}{\lambda + 1}$$

and on the other hand we have

$$
\begin{aligned}
F_2 &= \frac{\sum_{i=1}^{N-T^*} i^\gamma \big((i+T^*)/i\big)^\lambda \big(i/N\big)^\lambda}{\sum_{i=1}^{N-T^*} i^\gamma} - \frac{1}{\lambda+1} \\
&\geq \frac{1}{N^\lambda}\left(\frac{N}{N-T^*}\right)^\lambda \frac{\sum_{i=1}^{N-T^*} i^{\gamma+\lambda}}{\sum_{i=1}^{N-T^*} i^\gamma} - \frac{1}{\lambda+1} \\
&= \frac{1}{\left(N-T^*\right)^\lambda}\left(\frac{\left(N-T^*\right)^{1+\lambda+\gamma}}{\left(N-T^*\right)^{1+\gamma}}\frac{1+\gamma}{1+\lambda+\gamma} + o(1)\right) - \frac{1}{\lambda+1} \\
&= \frac{\lambda\gamma}{(1+\lambda+\gamma)(1+\lambda)} + o(1),
\end{aligned}
$$

hence the convergence is determined by F_1. A possible choice of N is $N = [\rho T^*]$ for some $1 < \rho < (\lambda+1)^{1/\lambda}$, which yields

$$
F_1 \simeq \frac{(\rho-1)^{1+\gamma}\,T^{*1+\gamma+\lambda}}{m^{1/2+\gamma+\lambda}\,h(\tilde G(\rho T^*/m))} \simeq \frac{(T^*/m)^{1+\gamma+\lambda}\,\sqrt{m}}{h(\tilde G(\rho T^*/m))}.
$$

For $h^{(1)}$ it holds that

$$
F_1 \simeq \left(\frac{T^*}{m}\right)^{\gamma-\lambda}\sqrt{m} \to \infty
$$

if $\gamma \geq \lambda$. For $h^{(2)}$ and $h^{(3)}$ it holds that

$$
F_1 \simeq \left(\frac{T^*}{m}\right)^{1/2+\gamma}\sqrt{m}/\log(T^*/m) \to \infty
$$

for arbitrary $\gamma, \lambda > 0$.

Note that in this section we assumed δ to be constant. Possible choices of N for $\delta_m \to 0$ can be found in Section 1.4.3.

Chapter 2

Monitoring gradual changes in renewal processes

In this chapter we establish results on detecting gradual changes in an unobservable (renewal) process based on observations of the corresponding counting process. Counting processes appear in many applications in various areas such as manufacturing, physics or insurance. For a general survey on counting processes and renewal theory we refer to Asmussen (2003) and Gut (2009) and for some application examples of counting processes we refer to Parzen (1999).

Gut and Steinebach (2002), (2009) considered the following scenario: One wishes to (sequentially) detect an (abrupt) change in the location of an unobservable process $\{S_k \,|\, k \in \mathbb{N}_0\}$, where S_k is the sum of k independently, and up to a possible shift in the mean identically distributed random variables, via observations of the corresponding counting process $N(t) := \inf\{k \,|\, S_k > t\}$. They suggest stopping times and detectors based on (equally weighted) increments of the observed process $N(t)$. Motivated by Gut (2011), who investigated the behavior of first passage times, with an underlying partial sum process that exhibits a non-linear trend, the question arises how to detect a gradual change in the above setting. As described earlier, if one expects to detect a gradual change, it is reasonable to put heavier weights on later observations where a (quasi) maximum likelihood approach suggests to choose a weight function based on the assumed type of change. The idea which we pursue in this chapter is to consider $\tilde{S}_k := \inf\{t \in \mathbb{N}_0 \,|\, N(t) > k\}$, i.e. the inverse process corresponding to our observed process $\{N(t) \,|\, t \in \mathbb{N}_0\}$, expecting that $\tilde{S}_k$ behaves similarly as S_k does. Hence, we construct our detectors as we would, if we observed $\{S_k\}$ directly, i.e. we consider weighted increments of the observations $\tilde{S}_k$, where the weight function is deduced from the (assumed) type of change.

This chapter is organized as follows: The testing problem is introduced in detail in Section 2.1. Suitable detectors and stopping times for closed-end and open-end monitoring procedures are introduced in Section 2.2. In Section 2.3 we demonstrate how a strong approximation of $\{S_n \,|\, n \in \mathbb{N}_0\}$ implies a strong approximation of $\{\tilde{S}_n \,|\, n \in \mathbb{N}_0\}$ under the null hypothesis and under the alternative. Making use of these results we can describe the asymptotic behavior of the stopping times under the null hypothesis (Section 2.4), show consistency of the suggested testing procedures (Section 2.5) and establish results on the asymptotic normality of the (suitably standardized) delay times (Section 2.6) under the alternative. Each section is again

61

subdivided with respect to known and unknown in-control parameters. The finite sample behavior is investigated in a small simulation study in Section 2.7.

2.1 Setting of the problem

Consider the following counting process

$$N(t) = \inf\{k \mid S_k > t\}, \quad t \geq 0,$$

sequentially observed at integer time points (i.e. we monitor $N(0)$, $N(1)$, $N(2), \dots$),
where $\{S_t \mid t \geq 0\}$ is an unobservable process that consists of a cumulated noise
term Y and a mean function M:

$$S_t = Y(t) + M(t).$$

More precisely $\{Y(t) \mid t \geq 0\}$ shall be a general stochastic process, which fulfills a
strong invariance principle

$$\sup_{0 \leq t \leq T} |Y(t) - \sigma W(t)| = \sup_{0 \leq t \leq T} |S_t - M(t) - \sigma W(t)| \stackrel{\text{a.s.}}{=} \mathcal{O}(T^\kappa) \qquad (2.1.1)$$

for some $\sigma > 0$, some $0 < \kappa < 1/2$ and some Wiener process $\{W(t) \mid t \geq 0\}$. (Examples for such processes are given in Section 1.1 and e.g. in Section 2.1 of Kirch
(2006).)

We are interested in testing for a possible change in the mean function M, which
we characterize via its increments denoted by $\mu(k) = M(k) - M(k-1)$. We assume that $\tilde{S}_k$ has a linear drift in the in-control state, i.e. constant increments
$M(k) - M(k-1) =: \mu_0 > 0$, which might start to in- or decrease (monotonously)
towards a (different) level $\mu_1 := \mu_0 + \delta > 0$ at some unknown time point k^*:

$$\mu(k) = \mu_1 - \delta \left(1 + (k - k^*)_+\right)^{-\gamma} = \mu_0 + \delta \left(1 - (1 + (k - k^*)_+)^{-\gamma}\right), \qquad (2.1.2)$$

where for the sake of simplicity we assume $k^* \in \mathbb{N}$. For the consistency results (see
Section 2.5) we also allow for local alternatives, i.e. $\delta = \delta_m \to 0$ as $m \to \infty$.

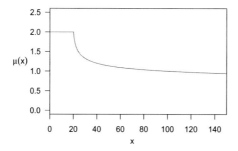

Figure 2.1: $\mu(x)$ for $\mu_0 = 2$, $\mu_1 = 1/2$, $\gamma = 1/4$ and $k^* = 20$

One may think for example of $N(t)$ as the number of insurance claims arriving in a certain time period. Calculations, like pricing insurance policies, can be based on the frequently collected counting data rather than on the (possibly unobservable) inter-arrival times and in our framework one may test whether the frequency of e.g. accidents increases.

We want to consider a closed-end as well as an open-end scenario. In both settings we rely on the non-contamination assumption, i.e. we assume that we have some training period of length

$$m < \mu_0\, k^*, \quad m \in \mathbb{N}, \tag{2.1.3}$$

during which $\mu(k)$ stays constant. In the closed-end setting we also have a fixed truncation point $T = T_m = [\vartheta\, m^\rho]$, where the sequential testing procedure will finally terminate, even if no change has been detected up to that time point. The truncation point shall either be a multiple of the training period, i.e. $\rho = 1$ and $\vartheta > 1$, or dominate the training period in the sense that $\rho > 1$ and $\vartheta \geq 1$. To shorten the notation we further introduce

$$\tilde{\vartheta} := \begin{cases} 1/\vartheta & \text{if } \rho = 1 \\ 0 & \text{if } \rho > 1. \end{cases}$$

In the open-end scenario we have no such fixed time horizon, but rather monitor the process until we either detect a change or (theoretically speaking) forever.

We are interested in testing the null hypothesis

$H_0: \delta = 0$ "no change"

against either one of the following closed-end alternatives

$H_{1,+}^{closed}: \delta > 0, \ k^* < T/\mu_0$ "one-sided, positive change",

$H_{1,-}^{closed}: \delta < 0, \ k^* < T/\mu_0$ "one-sided, negative change",

$H_1^{closed}: \delta \neq 0, \ k^* < T/\mu_0$ "two-sided change",

or one of the following open-end alternatives

$H_{1,+}^{open}: \delta > 0, \ k^* < \infty$ "one-sided, positive change",

$H_{1,-}^{open}: \delta < 0, \ k^* < \infty$ "one-sided, negative change",

$H_1^{open}: \delta \neq 0, \ k^* < \infty$ "two-sided change".

Remark 2.1.1. *The term "renewal process" usually refers to partial sums of independently, identically distributed random variables, which are almost sure nonnegative (e.g. inter-arrival times), whereas the more general case considered here is also known under the heading "renewal theory for random walks". Further, the term "counting process" rather refers to $N'(t) = \sup\{k\,|\,S_k \leq t\}$ than to $N(t) = \inf\{k\,|\,S_k > t\}$, which are known as "first passage times". If non-negative summands are assumed, $N(t) = N'(t) + 1$. For general summands, however, it is essential to consider the first passage times $N(t)$ which are stopping times with respect to the filtration $\{\mathcal{F}_n\}_{n \in \mathbb{N}}$, where $\mathcal{F}_n = \sigma\{S_k, k \leq n\}$ (cf. Section 3 of Gut (2011)).*

2.2 Stopping times

In this section we introduce detectors and stopping times for testing the null hypothesis of no change against either of the alternatives introduced in the previous section. In our setting we assume that a change might occur in a process $\{S_t \,|\, t \geq 0\}$, however, we only observe the corresponding counting process $N(t) = \inf\{k \,|\, S_k > t\}$. The observations are made at discrete time points, i.e. in the closed-end setting

$$t \in \{0, \ldots, m, \ldots, T\} =: \mathcal{I}_c$$

and in the open-end setting

$$t \in \{0, \ldots, m, \ldots, \infty\} =: \mathcal{I}_o.$$

Our test statistic will make use of the inverse of $N(t)$, that is

$$\tilde{S}_k = \inf\{t \in \mathcal{I} \,|\, N(t) > k\}, \quad k \in \mathbb{N}_0,$$

where $\mathcal{I}$ denotes either $\mathcal{I}_c$ or $\mathcal{I}_o$, depending on the setting we are working with.

The general idea is that $\{\tilde{S}_n \,|\, n \in \mathbb{N}_0\}$ behaves similarly to the unobservable realizations of $\{S_n \,|\, n \in \mathbb{N}_0\}$. This duality is further described in Section 2.3, where we resume how the invariance principle for $\{Y(t) \,|\, t \geq 0\}$ (or $\{S_t \,|\, t \geq 0\}$, respectively) can be converted into an invariance principle for $\{\tilde{S}_n \,|\, n \in \mathbb{N}_0\}$. One problem arising in this setting is that the number of observations of $\tilde{S}_n$ depends on $N(t)$, hence it is random. In Section 2.2.1 we give some asymptotics for the length of the training period m and the truncation point T. In Section 2.2.2 and Section 2.2.3 we propose detectors and stopping times for known or unknown in-control parameters, respectively. Note that in this setting we do not allow for a change in σ^2, so strictly speaking, σ^2 is not an in-control parameter. However, for the sake of clarity, we will still refer to μ_0 and σ^2 as such.

2.2.1 Randomness of the sample size

The sequential testing procedure shall start after the training period, that is after m observations of the counting process $N(t)$ are made. At this time point we have $\tilde{S}_k < \infty$ if and only if $k < N(m)$, hence we have $\tilde{m} = N(m) - 1$ observations of the process $\{\tilde{S}_k\}$. The asymptotic size of $N(t)$ is determined by the following strong law of large numbers for counting processes: Under the null hypothesis it holds that

$$\frac{N(t)}{t} \xrightarrow{\text{a.s.}} \frac{1}{\mu_0},$$

which is a well known result if S_n is e.g. the sum of independently, identically distributed random variables (cf. e.g. Theorem 3.4.1 of Gut (2009)) and, in our case, is an immediate consequence of Assertion (2.3.1) below. Hence, it holds that

$$\tilde{m} \overset{\text{a.s.}}{=} m/\mu_0 + o(m).$$

In the closed-end setting we also have to take the total number $\tilde{n}$ of observations of $\tilde{S}_n$ into account. By the same arguments as above we have that under the null hypothesis

$$\tilde{n} \stackrel{\text{a.s.}}{=} T/\mu_0 + o(T),$$

where T is the total number of observations of $N(t)$.

Under the alternative we do not have a precise asymptotic for $N(T)$ but rather asymptotic upper and lower bounds. In case of $\mu_1 < \mu_0$, i.e. $\delta < 0$, the bounds are derived as follows: Note that under the alternative we have $\mu_1 \leq \mu(k) \leq \mu_0$ for all k, hence

$$N_{\mu_0}(t) \leq N(t) \leq N_{\mu_1}(t),$$

where $N_{\mu_0}(t)$ and $N_{\mu_1}(t)$ are the counting processes based on $\{Y(n) + n\,\mu_0\}$ or $\{Y(n) + n\,\mu_1\}$, respectively. Similarly as above, we obtain

$$\frac{1}{\mu_0} \stackrel{\text{a.s.}}{\leq} \liminf_{T \to \infty} \frac{\tilde{n}}{T} \leq \limsup_{T \to \infty} \frac{\tilde{n}}{T} \stackrel{\text{a.s.}}{\leq} \frac{1}{\mu_1}. \tag{2.2.1}$$

Conversely, in case of $\mu_1 > \mu_0$ (2.2.1) holds true with $1/\mu_1$ as a lower and $1/\mu_0$ as an upper bound.

2.2.2 Stopping times for known in-control parameters

The monitoring procedure will be based on the weighted, standardized increments of the process $\tilde{S}_k$, i.e

$$R_k = \sum_{i=1}^{k} \frac{g(i)\left(\tilde{S}_i - \tilde{S}_{i-1} - \mu_0\right)}{\sigma}, \quad k \geq \tilde{m},$$

where $g(t) = 1 - (1+t)^{-\lambda}$ for some $\lambda > 0$. The choice of the weight function is, as in the previous chapter, motivated by a (quasi) maximum likelihood approach which suggests to use $(\mu(t) - \mu_0)/\delta$ as a weight function. Note that in our case the relation between the weight and the change function is $g(t) = \left(\mu(t + k^*) - \mu_0\right)/\delta$, if $\lambda = \gamma$.

On standardizing our detectors, we will make use of the function

$$G(t) = \int_0^t g^2(x)\,dx,$$

which will turn out to be asymptotically the variance of our detectors (see the corresponding discussion in Section 1.2.1).

We shall stop the monitoring procedure as soon as a detector (divided by a threshold function) exceeds some critical value c, thus our stopping times for the alternatives $H_{1,+}^{closed}$, $H_{1,-}^{closed}$ and H_1^{closed} are

$$\tau_{m,+}^{closed} = \min\left\{k = \tilde{m}, \ldots, \tilde{n} \mid R_k/h_1(k,T) > c\right\},$$
$$\tau_{m,-}^{closed} = \min\left\{k = \tilde{m}, \ldots, \tilde{n} \mid -R_k/h_1(k,T) > c\right\},$$
$$\tau_m^{closed} = \min\left\{k = \tilde{m}, \ldots, \tilde{n} \mid |R_k|/h_1(k,T) > c\right\},$$

whereas our stopping times for $H_{1,+}^{open}$, $H_{1,-}^{open}$ and H_1^{open} are

$$\tau_{m,+}^{open} = \min\left\{k = \tilde{m}, \ldots, \infty \mid R_k/h_2(k,\tilde{m}) > 1\right\},$$
$$\tau_{m,-}^{open} = \min\left\{k = \tilde{m}, \ldots, \infty \mid -R_k/h_2(k,\tilde{m}) > 1\right\},$$
$$\tau_m^{open} = \min\left\{k = \tilde{m}, \ldots, \infty \mid |R_k|/h_2(k,\tilde{m}) > 1\right\},$$

where h_1 and h_2 are suitably chosen threshold functions. It seems natural that the different settings (closed-end and open-end) require different threshold functions to bound the detectors. In the closed-end setting we will use

$$h_1(t,T) = G(t)^\beta G(T/\mu_0)^{1/2-\beta}, \quad 0 \leq \beta < 1/2,$$

where the parameter β controls the sensitivity of the testing procedure towards early or late changes. If an early change is expected, one would rather choose a large β, whereas if a late change is expected, one would rather choose a small β (also cf. Aue (2003), Sections 2.3 and 2.6).

In the open-end setting we consider the following threshold functions: For the one sided stopping times $\tau_{m,+}^{open}$ and $\tau_{m,-}^{open}$ we bound the detectors by

$$h_2(t, \tilde{m}) = \sqrt{G(t)}\; f^{-1}(\ln(G(t)/G(\tilde{m})) + f(c))\,,$$

with $f(t) = t^2 + 2\ln(\phi(t))$ and $c > 0$, and for the two-sided stopping time τ_m^{open} we bound the detectors by

$$h_2(t, \tilde{m}) = \sqrt{G(t)\;(\ln(G(t)/G(\tilde{m})) + c^2)},$$

with $c > 0$. These two threshold functions are chosen such that one can obtain critical values for the testing procedure via the distributions quoted in Remark 1.2.3 (see Theorem 2.4.2 and Theorem 2.4.6 for details).

2.2.3 Stopping times for unknown in-control parameters

In this section we suggest estimates for the typically unknown in-control parameters μ_0 and σ^2. The mean is estimated via

$$\hat{\mu}_{0,k} = \frac{1}{k}\tilde{S}_k,$$

whereas the variance σ^2 is replaced by a general estimate $\hat{\sigma}_k^2$ which fulfills under the null hypothesis

$$\sup_{k=\tilde{m},\ldots,\infty} \left|\hat{\sigma}_k^2 - \sigma^2\right| = o_P(1). \tag{2.2.2}$$

By replacing the unknown in-control parameters by the corresponding estimates we obtain the following detectors:

$$\hat{R}_k = \sum_{i=1}^{k} \frac{g(i)\left(\tilde{S}_i - \tilde{S}_{i-1} - \hat{\mu}_{0,k}\right)}{\hat{\sigma}_k}, \quad k \geq \tilde{m},$$

where $g(t) = 1 - (1+t)^{-\lambda}$ for some $0 < \lambda < 1/2 - \kappa$ (see the discussion on λ in Remark 2.2.1 below). Further, to shorten the notation we introduce $\Lambda := 1 - 2\lambda$.

Due to the estimate $\hat{\mu}_{0,k}$ the variance of the detectors slightly changes (see the corresponding discussion in Section 1.2.2). Hence, instead of standardizing the detectors by $G(t)$ we now have to standardize them by

$$\tilde{G}(t) = \int_0^t g^2(x)\,dx - \frac{\left(\int_0^t g(x)\,dx\right)^2}{t}. \tag{2.2.3}$$

We consider the following stopping times: for the alternatives $H_{1,+}^{closed}$, $H_{1,-}^{closed}$ and H_1^{closed}

$$\hat{\tau}_{m,+}^{closed} = \min\left\{k = \tilde{m},\ldots,\tilde{n} \,|\quad \hat{R}_k/\hat{h}_1(k,T) > c\right\},$$
$$\hat{\tau}_{m,-}^{closed} = \min\left\{k = \tilde{m},\ldots,\tilde{n} \,|\, -\hat{R}_k/\hat{h}_1(k,T) > c\right\},$$
$$\hat{\tau}_m^{closed} = \min\left\{k = \tilde{m},\ldots,\tilde{n} \,|\quad |\hat{R}_k|/\hat{h}_1(k,T) > c\right\}$$

and for the alternatives $H_{1,+}^{open}$, $H_{1,-}^{open}$ and H_1^{open}

$$\hat{\tau}_{m,+}^{open} = \min\left\{k = \tilde{m},\ldots,\infty \,|\quad \hat{R}_k/\hat{h}_2(k,\tilde{m}) > 1\right\},$$
$$\hat{\tau}_{m,-}^{open} = \min\left\{k = \tilde{m},\ldots,\infty \,|\, -\hat{R}_k/\hat{h}_2(k,\tilde{m}) > 1\right\},$$
$$\hat{\tau}_m^{open} = \min\left\{k = \tilde{m},\ldots,\infty \,|\quad |\hat{R}_k|/\hat{h}_2(k,\tilde{m}) > 1\right\}.$$

The corresponding threshold function for the closed-end setting is

$$\hat{h}_1(t,T) = \tilde{G}(t)^\beta \tilde{G}\left(T/\hat{\mu}_{0,[t]}\right)^{1/2-\beta}$$

and the corresponding threshold functions for the open-end setting are either for the one-sided stopping times $\hat{\tau}_{m,+}^{open}$ and $\hat{\tau}_{m,-}^{open}$,

$$\hat{h}_2(t, \tilde{m}) = \sqrt{\tilde{G}(t)} \, f^{-1}(\ln(\tilde{G}(t)/\tilde{G}(\tilde{m})) + f(c)),$$

with $f(t) = t^2 + 2\ln(\phi(t))$ and $c > 0$, or for the two-sided stopping time $\hat{\tau}_m^{open}$

$$\hat{h}_2(t, \tilde{m}) = \sqrt{\tilde{G}(t) \left(\ln(\tilde{G}(t)/\tilde{G}(\tilde{m})) + c^2 \right)}$$

with $c > 0$.

Remark 2.2.1. *The restriction on λ arises from the fact that (e.g. on proving the limit behavior of the testing procedure under the null hypothesis) we need $t^\kappa / \sqrt{\tilde{G}(t)} \to 0$, as $t \to \infty$. Since $\tilde{G}(t) \simeq t^\Lambda$ (cf. Section 2.2.4 below) this holds true for $\kappa < \Lambda/2$. Note that this assumption is not needed in case of known incontrol parameters, since $G(t) \simeq t$ (cf. Section 2.2.4), hence $t^\kappa / \sqrt{G(t)} \to 0$, as $t \to \infty$, is an immediate consequence of $\kappa < 1/2$.*

Remark 2.2.2. *A possible choice for an estimate fulfilling (2.2.2) under the null hypothesis is*

$$\hat{\sigma}_k^2 = \frac{1}{\hat{k}(k - \hat{k} + 1)} \sum_{j=\hat{k}}^{k} \left(\tilde{S}_j - \tilde{S}_{j-\hat{k}} - \hat{k} \, \tilde{S}_k/k \right)^2, \quad k \geq \tilde{m},$$

where $\hat{k} = k^q$ for some $2\kappa < q < 1$. (See Section 1.3.3; note that by Theorem 2.3.1 the assertion of Lemma 1.3.5 carries over in this setting if we replace $Z(k)$ by $\tilde{S}_k$.) In order to preserve the consistency under the alternative one can also estimate σ^2 non-sequentially, i.e. not with every newly made observation but rather by only using the data obtained in the training period.

2.2.4 Variance of the detectors

In this section we analyze the functions

$$G(t) = \int_0^t g^2(x)\,dx$$

and

$$\tilde{G}(t) = \int_0^t g^2(x)\,dx - \frac{\left(\int_0^t g(x)\,dx\right)^2}{t},$$

which were introduced in Section 2.2.2 and Section 2.2.3. Note that the function $g(t) = 1-(1+t)^{-\lambda}$ fulfills the assumptions of Lemma 1.2.7 (where for $\tilde{G}(t) \to \infty$ the restriction $\lambda < 1/2$ applies), hence the assertions of Lemma 1.2.7 hold true in the setting of this chapter, as well. The following additional growth rates will help us to prove the asymptotic results on the behavior of the test statistic.

Lemma 2.2.3. *Let* $g(t) = 1 - (1+t)^{-\lambda}$ *for some* $\lambda > 0$ *and* $\Lambda = 1 - 2\lambda$.

1. *It holds that* $G(t) \sim t$ *(as* $t \to \infty$), $\sup_{t \geq 1} t/G(t) < \infty$ *as well as* $\sup_{t \geq 1} G(t)/t < \infty$.

2. *If* $0 < \lambda < 1$ *we have* $\tilde{G}'(t) \sim \left(\lambda/\left((1-\lambda)t^\lambda\right)\right)^2$ *as* $t \to \infty$.

3. *If* $0 < \lambda < 1/2$ *we have*

$$\lim_{t \to \infty} \frac{\tilde{G}(t)}{t^\Lambda} = \frac{\lambda^2}{\Lambda\,(1-\lambda)^2} > 0,$$

$\sup_{t \geq 1} \tilde{G}(t)/t^\Lambda < \infty$ *and* $\sup_{t \geq 1} t^\Lambda/\tilde{G}(t) < \infty$.

Proof of Lemma 2.2.3. We prove the three assertions separately.

1. Integration yields

$$G(t) = t - 2\frac{(1+t)^{1-\lambda}-1}{1-\lambda} + \frac{(1+t)^\Lambda - 1}{\Lambda} \sim t,$$

   as $t \to \infty$. The latter two assertions follow by the continuity of $G(t)/t$.

2. Plugging the particular definition of $g(t)$ into the derivative calculated in Lemma 1.2.7 we obtain for $t > 0$

$$\begin{aligned}
\tilde{G}'(t) &= \left(\frac{g(t)\,t - \int_0^t g(x)\,dx}{t}\right)^2 \\
&= \left(\frac{-(1+t)^{-\lambda}t - \int_0^t -(1+x)^{-\lambda}\,dx}{t}\right)^2 \\
&= \left((1+t)^{-\lambda} - \frac{(1+t)^{1-\lambda}}{t\,(1-\lambda)} + \frac{1}{(1-\lambda)\,t}\right)^2
\end{aligned}$$

$$= t^{-2\lambda} \left(\left(\frac{1+t}{t} \right)^{-\lambda} - \frac{1}{1-\lambda} \left(\frac{1+t}{t} \right)^{1-\lambda} + \frac{1}{(1-\lambda)\, t^{1-\lambda}} \right)^2$$

$$\sim \left(\frac{\lambda}{(1-\lambda)\, t^{\lambda}} \right)^2.$$

3. Once again, integration yields for $t > 0$

$$\tilde{G}(t) = \frac{(1+t)^{\Lambda} - 1}{\Lambda} - \frac{1}{t} \left(\frac{(1+t)^{1-\lambda} - 1}{1-\lambda} \right)^2$$

and therefore, since $\lambda < 1/2$,

$$\frac{\tilde{G}(t)}{t^{\Lambda}} = \frac{(1+1/t)^{\Lambda} - 1/t^{\Lambda}}{\Lambda} - \left(\frac{(1+1/t)^{1-\lambda} - 1/t^{1-\lambda}}{1-\lambda} \right)^2$$

$$\to \frac{1}{\Lambda} - \frac{1}{(1-\lambda)^2}$$

$$= \frac{\lambda^2}{\Lambda\,(1-\lambda)^2} > 0.$$

The remaining two assertions follow by the continuity of $\tilde{G}(t)/t^{\Lambda}$.

$\square$

2.3 Preliminary results

Horváth (1986) has shown that under the null hypothesis the strong approximation for $\{S_t \,|\, t \geq 0\}$ (see (2.1.1)) implies a strong approximation for the corresponding inverse process $N(t) = \inf\{k \,|\, S_k > t\}$ by the same Wiener process with the best possible rate being $(t \log\log(t))^{1/4} \log(t)^{1/2}$. Csörgő et al. (1987) on the other hand, have constructed a different Wiener process which preserves the approximation rate of the underlying process. In Section 2.3.1 we apply their result to $\tilde{S}_n$ and obtain a strong approximation with the same rate as the strong approximation of S_n. In Section 2.3.2 we give an invariance principle for $\tilde{S}_n$ and (as a consequence) a growth rate for $\tilde{S}_n$, which holds true under the null hypothesis and under any of the alternatives introduced in Section 2.1.

2.3.1 Invariance principle under the null hypothesis

Theorem 2.3.1. *With the notation and assumptions of Section 2.1 it holds under the null hypothesis that, as $n \to \infty$,*

$$\sup_{0 \leq k \leq n} \left| \frac{\tilde{S}_k - k\,\mu_0}{\sigma} - \tilde{W}(k) \right| \overset{a.s.}{=} \mathcal{O}(n^\kappa),$$

where $\{\tilde{W}(t) \,|\, t \geq 0\}$ is a Wiener process.

Proof of Theorem 2.3.1. By Theorem 3.1 of Csörgő et al. (1987) we know that (2.1.1) implies that there exists a Wiener process $\{W^{(1)}(t) \,|\, t \geq 0\}$ such that

$$\sup_{0 \leq t \leq T} \left| \frac{N(t) - t/\mu_0}{\sigma/\mu_0^{3/2}} - W^{(1)}(t) \right| \overset{a.s.}{=} \mathcal{O}(T^\kappa), \tag{2.3.1}$$

where $N(t) = \inf\{k \,|\, S_k > t\}$. Denoting $Z(t) = N(t)\,\mu_0^{3/2}$ we have

$$\sup_{0 \leq t \leq T} \left| \frac{Z(t) - t\sqrt{\mu_0}}{\sigma} - W^{(1)}(t) \right| \overset{a.s.}{=} \mathcal{O}(T^\kappa).$$

Hence, by the same argument as above there is some Wiener process $\{W^{(2)}(t) \,|\, t \geq 0\}$ such that

$$\sup_{0 \leq t \leq T} \left| \frac{N^*(t) - t/\sqrt{\mu_0}}{\sigma/\mu_0^{3/4}} - W^{(2)}(t) \right| \overset{a.s.}{=} \mathcal{O}(T^\kappa),$$

where

$$
\begin{aligned}
N^*(t) &= \inf\{s \,|\, Z(s) > t\} \\
&= \inf\left\{ s \,|\, N(s) > t\,\mu_0^{-3/2} \right\} \\
&= \inf\left\{ k \in \mathbb{N} \,|\, N(k) > t\,\mu_0^{-3/2} \right\} + \mathcal{O}(1) \\
&= \tilde{S}_{t\mu_0^{-3/2}} + \mathcal{O}(1),
\end{aligned}
$$

where the transition from a continuous to a discrete argument follows by $N(t)$ being non-decreasing. Thus,

$$\sup_{0 \le t \le T} \left| \frac{N^*(t) - t/\sqrt{\mu_0}}{\sigma/\mu_0^{3/4}} - W^{(2)}(t) \right|$$

$$= \sup_{0 \le t \le T} \left| \frac{\tilde{S}_{t\mu_0^{-3/2}} - t/\sqrt{\mu_0}}{\sigma/\mu_0^{3/4}} - W^{(2)}(t) \right| + \mathcal{O}(1)$$

$$= \sup_{0 \le s \le S} \left| \frac{\tilde{S}_s - s\,\mu_0}{\sigma/\mu_0^{3/4}} - W^{(2)}\left(s\,\mu_0^{3/2}\right) \right| + \mathcal{O}(1),$$

where $S = \mu_0^{-3/2} T$. Setting $\tilde{W}(t) = \mu_0^{-3/4}\, W^{(2)}\left(t\,\mu_0^{3/2}\right)$ yields

$$\sup_{0 \le k \le n} \left| \frac{\tilde{S}_k - k\,\mu_0}{\sigma} - \tilde{W}(k) \right| \stackrel{\text{a.s.}}{=} \mathcal{O}(n^\kappa).$$

$\square$

2.3.2 Behavior under the alternative

The following two results hold true under the null hypothesis as well as under any of the alternatives introduced in Section 2.1. However, under the null hypothesis we can make use of the invariance principle of Theorem 2.3.1 which is certainly easier to handle. Thus, Theorem 2.3.2 can be seen as an analogue of Theorem 2.3.1 under the alternative.

Theorem 2.3.2. *With the notation and assumptions of Section 2.1 it holds that, as* $n \to \infty$,

$$\max_{k=1,\dots,n} \left| \tilde{S}_k - V(k) \right| \overset{a.s.}{=} \mathcal{O}(n^\kappa),$$

where

$$V(k) := \max_{l=1,\dots,k} \big(\sigma\, W(l) + M(l) \big), \tag{2.3.2}$$

with $\left\{ W(t) \,|\, t \geq 0 \right\}$, σ *and* κ *as in* (2.1.1).

Proof of Theorem 2.3.2. By the duality

$$\{N(t) > k\} = \{\max\{S_1, \dots, S_k\} \leq t\}$$

we have

$$\tilde{S}_k = \inf\{t \in \mathcal{I} \,|\, \max\{S_1, \dots, S_k\} \leq t\},$$

hence we can rewrite

$$\tilde{S}_k = \begin{cases} \big[\max\{S_1, \dots, S_k\}\big] + 1 & \text{if } \max\{S_1, \dots, S_k\} \notin \mathbb{N}, \\ \max\{S_1, \dots, S_k\} & \text{if } \max\{S_1, \dots, S_k\} \in \mathbb{N}. \end{cases}$$

This yields, as $n \to \infty$,

$$\max_{k=1,\dots,n} \left| \tilde{S}_k - V(k) \right|$$
$$\leq \max_{k=1,\dots,n} \max_{l=1,\dots,k} \left| S_l - \sigma W(l) - M(l) \right| + \mathcal{O}(1)$$
$$\overset{a.s.}{=} \mathcal{O}(n^\kappa).$$

$\square$

Via Theorem 2.3.2 one can show the following growth rate for $\tilde{S}_k$:

Corollary 2.3.3. *With the notation and assumptions of Section 2.1 it holds that, as* $n \to \infty$,

$$\max_{k=1,\dots,n} \left| \tilde{S}_k - M(k) \right| \overset{a.s.}{=} \mathcal{O}\big(\sqrt{n \log\log(n)}\big).$$

Proof of Corollary 2.3.3. By Theorem 2.3.2 and the law of the iterated logarithm we have, as $n \to \infty$,

$$
\begin{aligned}
&\max_{k=1,\ldots,n} \left| \tilde{S}_k - M(k) \right| \\
&\stackrel{\text{a.s.}}{=} \max_{k=1,\ldots,n} \left| V(k) - \max_{l=1,\ldots,k} M(l) \right| + \mathcal{O}(n^\kappa) \\
&\leq \max_{l=1,\ldots,n} \left| \sigma\, W(l) \right| + \mathcal{O}(n^\kappa) \\
&\stackrel{\text{a.s.}}{=} \mathcal{O}\left(\sqrt{n \, \log\log(n)} \right).
\end{aligned}
$$

$\square$

2.4 Asymptotics under the null hypothesis

In this section we establish results on the asymptotic behavior of the stopping times introduced in Section 2.2.2. Theorem 2.4.1 and Theorem 2.4.2 can be used to adjust the threshold functions, such that the testing procedures attain a prescribed level α asymptotically.

2.4.1 Asymptotics for known in-control parameters

Theorem 2.4.1. *With the notation and assumptions of Section 2.1 and Section 2.2 we have under the null hypothesis for any $c \in \mathbb{R}$*

$$\lim_{m \to \infty} P(\tau_{m,+}^{closed} = \infty) = P\left(\sup_{\tilde{\vartheta} \leq t \leq 1} W(t)/t^\beta \leq c \right),$$

$$\lim_{m \to \infty} P(\tau_{m,-}^{closed} = \infty) = P\left(\sup_{\tilde{\vartheta} \leq t \leq 1} W(t)/t^\beta \leq c \right),$$

$$\lim_{m \to \infty} P(\tau_m^{closed} = \infty) = P\left(\sup_{\tilde{\vartheta} \leq t \leq 1} |W(t)|/t^\beta \leq c \right).$$

Theorem 2.4.2. *With the notation and assumptions of Section 2.1 and Section 2.2 we have under the null hypothesis for any $c > 0$*

$$\lim_{m \to \infty} P(\tau_{m,+}^{open} = \infty) = P\left(\sup_{t \geq 1} \frac{W(t)}{\sqrt{t}\ f^{-1}(\ln(t) + f(c))} \leq 1 \right),$$

$$\lim_{m \to \infty} P(\tau_{m,-}^{open} = \infty) = P\left(\sup_{t \geq 1} \frac{W(t)}{\sqrt{t}\ f^{-1}(\ln(t) + f(c))} \leq 1 \right),$$

$$\lim_{m \to \infty} P(\tau_m^{open} = \infty) = P\left(\sup_{t \geq 1} \frac{|W(t)|}{\sqrt{t}\ (\ln(t) + c^2)} \leq 1 \right).$$

Before we turn to the proofs of the theorems, we state two remarks which ensure the applicability of the results in practice.

Remark 2.4.3. *All of the random variables appearing on the right hand sides in the two theorems above have a non-degenerated limit distribution (cf. Lemma 2.1 in Section 4.2 of Csörgő and Horváth (1993), Theorem 1.5.1 of Csörgő and Révész (1981) and Example 2 and Example 3 of Robbins and Siegmund (1970)). The distributions of $\sup_{\tilde{\vartheta} \leq t \leq 1} W(t)/t^\beta$ and $\sup_{\tilde{\vartheta} \leq t \leq 1} |W(t)|/t^\beta$ are not explicitly known for $\beta, \tilde{\vartheta} \neq 0$, thus they have to be simulated in applications. Some selected simulated critical values are given by Horváth et al. (2004). The distributions of the random variables on the right hand sides in Theorem 2.4.2 are known (see Remark 1.2.3).*

Remark 2.4.4. *By choosing the constant c we can adjust the threshold functions such that the probabilities on the right hand sides equal to $1 - \alpha$. In Theorem 2.4.2 we are limited to the case of $c > 0$. However, this restriction does not interfere in practice since it is fulfilled, if the asymptotic false alarm rate $\alpha \in (0, 1/2 + 1/\pi)$ (for $\tau_{m,+}^{open}$ and $\tau_{m,-}^{open}$,) or $\alpha \in (0, 1)$ (for τ_m^{open}), respectively (see Remark 1.2.3).*

Proof of Theorem 2.4.1. We make use of Remark 1.3.3, Lemma 2.2.3 and Theorem 2.3.1 to approximate the difference between our test statistic and the corresponding "Gaussian analogue":

$$
\begin{aligned}
&\max_{k=\tilde{m},\dots,\tilde{n}} \frac{\left| R_k - \sum_{i=1}^{k} g(i)(W(i) - W(i-1)) \right|}{h_1(k,T)} \\
&\leq \max_{k=\tilde{m},\dots,\tilde{n}} \frac{4\,g(k) \max_{l=1,\dots,k} \left| (\tilde{S}_l - \mu_0 l)/\sigma - W(l) \right|}{h_1(k,T)} \\
&\overset{a.s.}{=} \max_{k=\tilde{m},\dots,\tilde{n}} \frac{k^{\kappa}}{h_1(k,T)}\, \mathcal{O}(1) \\
&= \max_{k=\tilde{m},\dots,\tilde{n}} \frac{k^{\kappa}}{k^{\beta}\, T^{1/2-\beta}} \left(\frac{k}{G(k)} \right)^{\beta} \left(\frac{T/\mu_0}{G(T/\mu_0)} \right)^{\frac{1}{2}-\beta} \mathcal{O}(1) \\
&\overset{a.s.}{=} \max_{k=\tilde{m},\dots,\tilde{n}} (k/T)^{\frac{1}{2}-\beta}\, k^{\kappa-1/2}\, \mathcal{O}(1) \\
&= \tilde{m}^{\kappa-1/2}\, \mathcal{O}(1) \\
&\overset{a.s.}{=} o(1).
\end{aligned}
$$
(2.4.1)

Next, we approximate the sum of the weighted increments of $\{W(t)\,|\,t \geq 0\}$ by a Wiener integral:

$$
\begin{aligned}
&\sum_{i=1}^{k} g(i)\,(W(i) - W((i-1))) \\
&\overset{a.s.}{=} g(k)\,W(k) - \sum_{i=0}^{k-1} (g(i+1) - g(i))\,W(i) \\
&= g(k)\,W(k) - \sum_{i=0}^{k-1} \int_{i}^{i+1} g'(x)\,W(i)\,dx \\
&= \int_{0}^{k} g(x)\,dW(x) + \sum_{i=0}^{k-1} \int_{i}^{i+1} g'(x)(W(i) - W(x))\,dx,
\end{aligned}
$$

where by Proposition 1.3.2 it holds that

$$
\begin{aligned}
&\max_{k=\tilde{m},\dots,\tilde{n}} \left| \sum_{i=0}^{k-1} \int_{i}^{i+1} \frac{g'(x)(W(i) - W(x))}{h_1(k,T)}\,dx \right| \\
&\leq \max_{k=\tilde{m},\dots,\tilde{n}} \max_{i=0,\dots,k-1} \sup_{i \leq x \leq i+1} |W(i) - W(x)| \frac{\int_{0}^{k} g'(x)\,dx}{h_1(k,T)} \\
&\leq \max_{0 \leq z \leq \tilde{n}-1} \max_{0 \leq \epsilon \leq 1} |W(z) - W(z+\epsilon)| \max_{k=\tilde{m},\dots,\tilde{n}} \frac{g(k)}{h_1(k,T)} \\
&\overset{a.s.}{=} \mathcal{O}(\sqrt{\log(T)}) \max_{k=\tilde{m},\dots,\tilde{n}} \frac{(G(T/\mu_0)/G(k))^{\beta}}{\sqrt{G(T/\mu_0)}} \\
&= \mathcal{O}(\sqrt{\log(T)}\, T^{\beta-\frac{1}{2}}) \\
&= o(1).
\end{aligned}
$$
(2.4.2)

Hence, it holds that

$$\sup_{k=\tilde{m},\dots,\tilde{n}} \frac{R_k}{h_1(k,T)} \overset{a.s.}{=} \sup_{k=\tilde{m},\dots,\tilde{n}} \frac{\int_0^k g(x)\,dW(x)}{h_1(k,T)} + o(1). \tag{2.4.3}$$

Analyzing the covariance structure of the processes (see (1.3.6)) yields

$$\left\{ \int_0^t g(x)\,dW(x) \,\big|\, t \ge 0 \right\} \overset{D}{=} \{ W(G(t)) \,|\, t \ge 0 \}. \tag{2.4.4}$$

The next step is a transition to a continuous argument:

$$\begin{aligned}
\sup_{\tilde{m} \le t \le \tilde{n}} & \left| \frac{W(G([t]))}{h_1([t],T)} - \frac{W(G(t))}{h_1(t,T)} \right| \\
&\le \sup_{\tilde{m} \le t \le \tilde{n}} \frac{|W(G([t]))|}{h_1([t],T)} \left(1 - \frac{h_1([t],T)}{h_1(t,T)} \right) \\
&\quad + \sup_{\tilde{m} \le t \le \tilde{n}} \frac{|W(G([t])) - W(G(t))|}{h_1(t,T)} \\
&= o_P(1),
\end{aligned} \tag{2.4.5}$$

where we already plugged in the following three asymptotic relations:

1. As we proceed in the proof, we see that

$$\sup_{\tilde{m} \le t \le \tilde{n}} \frac{|W(G([t]))|}{h_1([t],T)} = \mathcal{O}_P(1).$$

2. If $\beta = 0$, we have $1 - h_1([t],T)/h_1(t,T) = 0$. If $\beta > 0$, we have by the mean value theorem

$$\begin{aligned}
\sup_{\tilde{m} \le t \le \tilde{n}} & \left(1 - \frac{h_1([t],T)}{h_1(t,T)} \right) \\
&= \sup_{\tilde{m} \le t \le \tilde{n}} \frac{G^\beta(t) - G^\beta([t])}{G^\beta(t)} \\
&= \sup_{\tilde{m} \le t \le \tilde{n}} \beta \frac{G^{\beta-1}(\xi)\,g^2(\xi)}{G^\beta(t)} \quad \text{for some } \xi \in [[t],t] \\
&\le \sup_{\tilde{m} \le t \le \tilde{n}} \beta \left(\frac{G(\xi)}{G(t)} \right)^\beta \frac{1}{G([t])} \\
&\overset{a.s.}{=} \mathcal{O}(1/m) \\
&= o(1).
\end{aligned} \tag{2.4.6}$$

3. Since $0 \le G(t) - G([t]) = \int_{[t]}^t g^2(x)\,dx \le 1$, we have by Proposition 1.3.2 and

Lemma 2.2.3 that

$$\sup_{\tilde{m}\leq t\leq \tilde{n}} \frac{|W(G(t)) - W(G([t]))|}{h_1(t,T)}$$

$$\overset{a.s.}{=} \sup_{\tilde{m}\leq t\leq \tilde{n}} \frac{\sqrt{\log(T)}}{t^{\beta}\, T^{1/2-\beta}}\, \mathcal{O}(1) \tag{2.4.7}$$

$$= \sqrt{\log(T)}\, \tilde{m}^{-(\beta+\rho(1/2-\beta))}\, \mathcal{O}(1)$$

$$\overset{a.s.}{=} o(1),$$

where the last rate of convergence follows by $\beta < 1/2$.

On setting $b_1 = G(\tilde{m})/G(T/\mu_0)$ and $b_2 = G(\tilde{n})/G(T/\mu_0)$ we have

$$\sup_{\tilde{m}\leq t\leq \tilde{n}} \frac{W(G(t))}{h_1(t,T)} = \sup_{\tilde{m}\leq t\leq \tilde{n}} \frac{W(G(t))/\sqrt{G(T/\mu_0)}}{(G(t)/(G(T/\mu_0)))^{\beta}} \overset{D}{=} \sup_{b_1\leq t\leq b_2} \frac{W(t)}{t^{\beta}}. \tag{2.4.8}$$

Note that by Lemma 2.2.3 and $T = [\vartheta\, m^{\rho}]$ we have

$$b_1 = \frac{G(\tilde{m})}{G(T/\mu_0)} = \frac{G(\tilde{m})}{\tilde{m}}\, \frac{T/\mu_0}{G(T/\mu_0)}\, \frac{\tilde{m}}{T/\mu_0} \overset{a.s.}{\longrightarrow} \tilde{\vartheta} = \begin{cases} 1/\vartheta & \text{if } \rho = 1, \\ 0 & \text{if } \rho > 1, \end{cases}$$

and

$$b_2 = \frac{G(\tilde{n})}{\tilde{n}}\, \frac{T/\mu_0}{G(T/\mu_0)}\, \frac{\tilde{n}}{T/\mu_0} \overset{a.s.}{\longrightarrow} 1.$$

Thus, we have by the almost sure continuity of $W(t)/t^{\beta}$

$$\sup_{b_1\leq t\leq b_2} \frac{W(t)}{t^{\beta}} \overset{a.s.}{\longrightarrow} \sup_{\tilde{\vartheta}\leq t\leq 1} \frac{W(t)}{t^{\beta}}. \tag{2.4.9}$$

Combining (2.4.3), (2.4.4), (2.4.5), (2.4.8) and (2.4.9) yields

$$\sup_{k=\tilde{m},\dots,\tilde{n}} \frac{R_k}{h_1(k,T)} \overset{D}{\longrightarrow} \sup_{\tilde{\vartheta}\leq t\leq 1} \frac{W(t)}{t^{\beta}}, \tag{2.4.10}$$

which yields the first assertion of Theorem 2.4.1. The latter two assertions follow by similar arguments. $\qquad\qquad\Box$

Proof of Theorem 2.4.2. By Remark 1.7.1 we know that, as $t \to \infty$,

$$\sqrt{t}\, f^{-1}(\ln(t) + f(c)) \sim \sqrt{t\, (\ln(t) + c^2)},$$

hence if we show the assertion for the stopping time $\tau_{m,+}^{open}$, the proof carries over to the assertions for the stopping times $\tau_{m,-}^{open}$ and τ_m^{open}.

Much of the proof follows along the lines of the proof of Theorem 2.4.1. As usual, the first step is the transition to a Gaussian version of our test statistic. Along the lines of (2.4.1) we have by Lemma 2.2.3

$$
\max_{k=\tilde{m},\dots,\infty} \frac{\left|R_k - \sum_{i=1}^{k} g(i)(W(i) - W(i-1))\right|}{h_2(k,\tilde{m})}
$$
$$
\stackrel{\text{a.s.}}{=} \max_{k=\tilde{m},\dots,\infty} \frac{g(k)\, k^{\kappa}}{h_2(k,\tilde{m})}\, \mathcal{O}(1)
$$
$$
\stackrel{\text{a.s.}}{=} \frac{\tilde{m}^{\kappa}}{\sqrt{G(\tilde{m})}}\, \mathcal{O}(1)
$$
$$
\stackrel{\text{a.s.}}{=} o(1).
$$
(2.4.11)

Following (2.4.2) we find that

$$
\max_{k=\tilde{m},\dots,\infty} \frac{\left|\sum_{i=1}^{k} g(i)(W(i) - W(i-1)) - \int_0^k g(x)\, dW(x)\right|}{h_2(k,\tilde{m})}
$$
$$
\stackrel{\text{a.s.}}{=} \mathcal{O}(1) \max_{k=\tilde{m},\dots,\infty} \frac{g(k)\, \sqrt{\log(k)}}{h_2(k,\tilde{m})}
$$
$$
\stackrel{\text{a.s.}}{=} o(1).
$$
(2.4.12)

Hence, we can make use of (2.4.4) again. As in (2.4.5) we see that

$$
\sup_{t \geq \tilde{m}} \left| \frac{h_2(t,\tilde{m})}{h_2([t],\tilde{m})} - 1 \right| \stackrel{\text{a.s.}}{=} o(1)
$$
(2.4.13)

and

$$
\sup_{t \geq \tilde{m}} \frac{|W(G(t)) - W(G([t]))|}{h_2(t,\tilde{m})} \stackrel{\text{a.s.}}{=} o(1),
$$
(2.4.14)

yield the transition to a time continuous process, i.e. (if (2.4.13) and (2.4.14) hold true) we have

$$
\sup_{k=\tilde{m},\dots,\infty} \frac{W(G(k))}{h_2(k,\tilde{m})} = \sup_{t \geq \tilde{m}} \frac{W(G(t))}{h_2(t,\tilde{m})} + o_P(1).
$$

Further, by Lemma 2.2.3 we then obtain

$$
\sup_{t \geq \tilde{m}} \frac{W(G(t))}{h_2(t,\tilde{m})}
$$
$$
\stackrel{D}{=} \sup_{t \geq \tilde{m}} \frac{W\big(G(t)/G(m/\mu_0)\big)}{\sqrt{G(t)/G(m/\mu_0)}\; f^{-1}\big(\ln(G(t)/G(\tilde{m})) + f(c)\big)}
$$
$$
\stackrel{\text{a.s.}}{\longrightarrow} \sup_{t \geq 1} \frac{W(t)}{\sqrt{t}\; f^{-1}(\ln(t) + f(c))},
$$
(2.4.15)

which shows the first assertion of Theorem 2.4.2 and, as before, the second and third assertion follow by the same arguments. Thus, we focus on proving (2.4.13) and

(2.4.14): Assertion (2.4.14) can easily be verified: Via Proposition 1.3.2 and Lemma 2.2.3 we have

$$\sup_{t \geq \tilde{m}} \frac{|W(G(t)) - W(G([t]))|}{h_2(t, \tilde{m})} \overset{\text{a.s.}}{=} \sup_{t \geq \tilde{m}} \frac{\sqrt{\log(t)}}{\sqrt{G(t)}} \mathcal{O}(1) \overset{\text{a.s.}}{=} o(1).$$

Proving Assertion (2.4.13) is somewhat more tedious: On rewriting

$$0 \leq \frac{h_2(t, \tilde{m})}{h_2([t], \tilde{m})} - 1 = \sqrt{\frac{G(t)}{G([t])} \frac{f^{-1}(\ln(G(t)/G(\tilde{m})) + f(c))}{f^{-1}(\ln(G([t])/G(\tilde{m})) + f(c))}} - 1$$

we see that, on the one hand,

$$1 \leq \sup_{t \geq \tilde{m}} \frac{G(t)}{G([t])} = \sup_{t \geq \tilde{m}} \frac{G(t) - G([t])}{G([t])} + 1 \leq \frac{1}{G(\tilde{m})} + 1 \overset{\text{a.s.}}{=} 1 + \mathcal{O}(m^{-1})$$

and, on the other hand, it holds by the mean value theorem for some $\xi \in [G([t])/G(\tilde{m}) + f(c), G(t)/G(\tilde{m}) + f(c)]$ that

$$\begin{aligned}
1 &\leq \frac{f^{-1}(\ln(G(t)/G(\tilde{m})) + f(c))}{f^{-1}(\ln(G([t])/G(\tilde{m})) + f(c))} \\
&= \frac{f^{-1\prime}(\xi)}{f^{-1}(\ln(G([t])/G(\tilde{m})) + f(c))} \ln\left(\frac{G(t)/G(\tilde{m}) + f(c)}{G([t])/G(\tilde{m}) + f(c)}\right) + 1 \\
&\leq \frac{f^{-1\prime}(\xi)}{c} \ln\left(1 + \frac{G(t) - G(G([t]))}{G([t]) + G(\tilde{m}) f(c)}\right) + 1 \\
&\leq \frac{f^{-1\prime}(\xi)}{c} \ln\left(1 + \frac{1}{G([t]) + G(\tilde{m}) f(c)}\right) + 1.
\end{aligned}$$

By Lemma 2.2.3 we have

$$\sup_{t \geq \tilde{m}} \ln\left(1 + \frac{1}{G([t]) + G(\tilde{m}) f(c)}\right) \overset{\text{a.s.}}{=} o(1)$$

and since $f'(t) = 2t + 2\varphi(t)/\phi(t) \geq 2t$ it holds that

$$\sup_{\xi \geq f(c)} f^{-1\prime}(\xi) = \sup_{\xi \geq f(c)} \frac{1}{f'(f^{-1}(\xi))} \leq \sup_{\xi \geq f(c)} \frac{1}{2 f^{-1}(\xi)} \leq \frac{1}{2c}. \tag{2.4.16}$$

Combining the assertions above yields (2.4.13), which completes the proof. □

2.4.2 Asymptotics for unknown in-control parameters

Theorem 2.4.5. *With the notation and assumptions of Section 2.1 and Section 2.2 we have under the null hypothesis for any* $c \in \mathbb{R}$

$$\lim_{m \to \infty} P\big(\hat{\tau}_{m,+}^{closed} = \infty\big) = P\bigg(\sup_{\tilde{\vartheta} \leq t \leq 1} W(t)/t^\beta \leq c \bigg),$$

$$\lim_{m \to \infty} P\big(\hat{\tau}_{m,-}^{closed} = \infty\big) = P\bigg(\sup_{\tilde{\vartheta} \leq t \leq 1} W(t)/t^\beta \leq c \bigg),$$

$$\lim_{m \to \infty} P\big(\hat{\tau}_{m}^{closed} = \infty\big) = P\bigg(\sup_{\tilde{\vartheta} \leq t \leq 1} |W(t)|/t^\beta \leq c \bigg).$$

Theorem 2.4.6. *With the notation and assumptions of Section 2.1 and Section 2.2 we have under the null hypothesis for any* $c > 0$

$$\lim_{m \to \infty} P\big(\hat{\tau}_{m,+}^{open} = \infty\big) = P\bigg(\sup_{t \geq 1} \frac{W(t)}{\sqrt{t}\ f^{-1}(\ln(t) + f(c))} \leq 1 \bigg),$$

$$\lim_{m \to \infty} P\big(\hat{\tau}_{m,-}^{open} = \infty\big) = P\bigg(\sup_{t \geq 1} \frac{W(t)}{\sqrt{t}\ f^{-1}(\ln(t) + f(c))} \leq 1 \bigg),$$

$$\lim_{m \to \infty} P\big(\hat{\tau}_{m}^{open} = \infty\big) = P\bigg(\sup_{t \geq 1} \frac{|W(t)|}{\sqrt{t}\ (\ln(t) + c^2)} \leq 1 \bigg).$$

Proof of Theorem 2.4.5. We start by eliminating the estimate $\hat{\mu}_{0,k}$ in the threshold function: Denoting by

$$\tilde{h}_1(t,T) := \tilde{G}(t)^\beta \tilde{G}(T/\mu_0)^{\frac{1}{2}-\beta}$$

the function corresponding to $\hat{h}_1$, yet with the estimate $\hat{\mu}_{0,t}$ being replaced by the true value μ_0, we have

$$\max_{k=\tilde{m},\dots,\tilde{n}} \left| \frac{\hat{R}_k}{\hat{h}_1(k,T)} - \frac{\hat{R}_k}{\tilde{h}_1(k,T)} \right| \qquad (2.4.17)$$
$$\leq \max_{k=\tilde{m},\dots,\tilde{n}} \frac{|\hat{R}_k|}{\tilde{h}_1(k,T)} \max_{k=\tilde{m},\dots,\tilde{n}} \left| \bigg(\frac{\tilde{G}(T/\mu_0)}{\tilde{G}(T/\hat{\mu}_{0,k})} \bigg)^{\frac{1}{2}-\beta} - 1 \right|.$$

As we will see later on in this proof, the first factor is of order $\mathcal{O}_P(1)$ as $m \to \infty$. We take care of the second factor: By the mean value theorem we have

$$\left| \frac{\tilde{G}(T/\mu_0)}{\tilde{G}(T/\hat{\mu}_{0,k})} - 1 \right| = \left| \frac{\tilde{G}(T/\mu_0) - \tilde{G}(T/\hat{\mu}_{0,k})}{\tilde{G}(T/\hat{\mu}_{0,k})} \right| = \frac{\tilde{G}'(\xi)}{\tilde{G}(T/\hat{\mu}_{0,k})} \left| \frac{T}{\mu_0} - \frac{T}{\hat{\mu}_{0,k}} \right|$$

where $\xi = \xi_k \in [T/\mu_0, T/\hat{\mu}_{0,k}]$ or $\xi \in [T/\hat{\mu}_{0,k}, T/\mu_0]$, respectively. Hence, by Lemma 2.2.3 we have

$$\max_{k=\tilde{m},\dots,\tilde{n}} \left| \frac{\tilde{G}(T/\mu_0)}{\tilde{G}(T/\hat{\mu}_{0,k})} - 1 \right|$$
$$= \max_{k=\tilde{m},\dots,\tilde{n}} \frac{\xi^{-2\lambda}\ T/\hat{\mu}_{0,k}}{(T/\hat{\mu}_{0,k})^{1-2\lambda}} \left| \frac{\hat{\mu}_{0,k}}{\mu_0} - 1 \right| \mathcal{O}(1) \qquad (2.4.18)$$
$$\overset{\text{a.s.}}{=} o(1),$$

where for the last equality we used

$$\max_{k=\tilde{m},\dots,\tilde{n}} |\hat{\mu}_{0,k} - \mu_0| \overset{\text{a.s.}}{=} \mathcal{O}\big(\sqrt{\log\log(m)/m}\big) = o(1),$$

which is a consequence of Theorem 2.3.1 and the law of the iterated logarithm. Plugging this in (2.4.17) yields

$$\sup_{k=\tilde{m},\dots,\tilde{n}} \left| \frac{\hat{R}_k}{\tilde{h}_1(k,T)} - \frac{\hat{R}_k}{\tilde{h}_1(k,T)} \right| = o_P(1).$$

In addition, note that by the same arguments as in (1.3.7) we can neglect the estimate $\hat{\sigma}_k^2$ in our further considerations, thus we focus on determining the limit distribution of $\max_{k=\tilde{m},\dots,\tilde{n}} \tilde{R}_k/\tilde{h}_1(k,T)$, where $\tilde{R}_k := (\hat{\sigma}_k/\sigma)\,\hat{R}_k$. Decomposing

$$\frac{\tilde{R}_k}{\tilde{h}_1(k,T)} = \frac{R_k}{\tilde{h}_1(k,T)} + \frac{(\mu_0 - \hat{\mu}_{0,k})\sum_{i=1}^k g(i)}{\sigma\,\tilde{h}_1(k,T)}$$

we see that if

$$\sup_{k=\tilde{m},\dots,\tilde{n}} \left| \frac{R_k}{\tilde{h}_1(k,T)} - \frac{\int_0^k g(x)\,dW(x)}{\tilde{h}_1(k,T)} \right| \overset{\text{a.s.}}{=} o(1) \tag{2.4.19}$$

and

$$\sup_{k=\tilde{m},\dots,\tilde{n}} \left| \frac{(\mu_0 - \hat{\mu}_{0,k})\sum_{i=1}^k g(i)}{\sigma\,\tilde{h}_1(k,T)} - \frac{\int_0^k g(x)\,dx\,W(k)/k}{\tilde{h}_1(k,T)} \right| \overset{\text{a.s.}}{=} o(1) \tag{2.4.20}$$

hold true, we have

$$\sup_{k=\tilde{m},\dots,\tilde{n}} \left| \frac{\tilde{R}_k}{\tilde{h}_1(k,T)} - \frac{U(k)}{\tilde{h}_1(k,T)} \right| \overset{\text{a.s.}}{=} o(1),$$

where

$$U(t) = \begin{cases} \displaystyle\int_0^t \left(g(x) - \frac{\int_0^t g(y)\,dy}{t} \right) dW(x) & \text{for } t > 0, \\ 0 & \text{for } t = 0. \end{cases} \tag{2.4.21}$$

Analyzing the covariance structure as in (1.3.19) and taking the fact that $\{U(t)\,|\,t \geq 0\}$ is a Gaussian process into account gives us

$$\{U(t)\,|\,\tilde{m} \leq t \leq \tilde{n}\} \overset{D}{=} \{W\big(\tilde{G}(t)\big)\,|\,\tilde{m} \leq t \leq \tilde{n}\}. \tag{2.4.22}$$

So, if (2.4.19), (2.4.20) and, in addition,

$$\sup_{\tilde{m} \leq t \leq \tilde{n}} \left| \frac{W(\tilde{G}([t]))}{\tilde{h}_1([t],T)} - \frac{W(\tilde{G}(t))}{\tilde{h}_1(t,T)} \right| = o_P(1) \tag{2.4.23}$$

hold true, we have similarly as in (2.4.10), that

$$\sup_{k=\tilde{m},\ldots,\tilde{n}} \frac{\tilde{R}_k}{\tilde{h}_1(k,T)} \xrightarrow{D} \sup_{\tilde{\vartheta}\leq t\leq 1} \frac{W(t)}{t^{\beta}}, \tag{2.4.24}$$

which then completes the proof for the one-sided stopping time $\tau_{m,+}^{closed}$.

(2.4.19) follows along the lines of Theorem 1.3.1, yet we have to check that the convergence rates hold true for $\tilde{h}_1(k,T)$ instead of $h_1(k,T)$. Recall that $\Lambda = 1 - 2\lambda > 0$. Concerning (2.4.1) we note that

$$\max_{k=\tilde{m},\ldots,\tilde{n}} \frac{g(k)\,k^{\kappa}}{\tilde{h}_1(k,T)} \overset{a.s.}{=} \max_{k=\tilde{m},\ldots,\tilde{n}} k^{\kappa-\Lambda/2} \left(\frac{k}{T}\right)^{\Lambda(1/2-\beta)} \mathcal{O}(1) \overset{a.s.}{=} o(1)$$

by our assumptions on λ. As to (2.4.2) we have

$$\begin{aligned}
\sqrt{\log(T)} &\max_{k=\tilde{m},\ldots,\tilde{n}} \frac{g(k)}{\tilde{h}_1(k,T)} \\
&= \sqrt{\log(T)} \max_{k=\tilde{m},\ldots,\tilde{n}} k^{-\Lambda/2} \left(\frac{k}{T}\right)^{\Lambda(1/2-\beta)} \mathcal{O}(1) \\
&\overset{a.s.}{=} \sqrt{\log(T)}\, m^{-\Lambda/2}\, \mathcal{O}(1) \\
&= o(1)
\end{aligned}$$

which shows (2.4.19). As to (2.4.20) we have by the invariance principle of Theorem 2.3.1

$$\begin{aligned}
\max_{k=\tilde{m},\ldots,\tilde{n}} &\left| \frac{(\mu_0 - \hat{\mu}_{0,k} + \sigma\,W(k)/k)\sum_{i=1}^{k} g(i)}{\sigma\,\tilde{h}_1(k,T)} \right| \\
&\overset{a.s.}{=} \max_{k=\tilde{m},\ldots,\tilde{n}} \frac{k^{\kappa}}{\tilde{h}_1(k,T)} \frac{\sum_{i=1}^{k} g(i)}{k} \mathcal{O}(1) \\
&\overset{a.s.}{=} o(1)
\end{aligned} \tag{2.4.25}$$

and further

$$\begin{aligned}
\sup_{k=\tilde{m},\ldots,\tilde{n}} &\left| \frac{\sum_{i=1}^{k} g(i)W(k)/k}{\tilde{h}_1(k,T)} - \frac{\int_0^k g(x)\,dx\,W(k)/k}{\tilde{h}_1(k,T)} \right| \\
&= \sup_{k=\tilde{m},\ldots,\tilde{n}} \left| \frac{W(k)}{\tilde{h}_1(k,T)\,k} \left(\sum_{i=1}^{k} g(i) - \int_0^k g(x)\,dx \right) \right|,
\end{aligned} \tag{2.4.26}$$

where, on the one hand, we have

$$\sup_{k=\tilde{m},\ldots,\tilde{n}} \left| \frac{W(k)}{\tilde{h}_1(k,T)\,k} \right| \overset{a.s.}{=} \sup_{k=\tilde{m},\ldots,\tilde{n}} \frac{\sqrt{k\,\log\log(k)}}{k^{\Lambda/2}\,k} \mathcal{O}(1) \overset{a.s.}{=} o(1)$$

and, on the other hand, we have uniformly for $\tilde{m} \leq k \leq \tilde{n}$

$$\left| \sum_{i=1}^{k} g(i) - \int_{0}^{k} g(x)\, dx \right| \leq \sum_{i=1}^{k} g(i) - g(i-1) = \mathcal{O}(1),$$

which yields (2.4.20). Concerning (2.4.23) we decompose

$$\sup_{\tilde{m} \leq t \leq \tilde{n}} \left| \frac{W(\breve{G}([t]))}{\tilde{h}_1([t], T)} - \frac{W(\breve{G}(t))}{\tilde{h}_1(t, T)} \right|$$

$$\leq \sup_{\tilde{m} \leq t \leq \tilde{n}} \frac{|W(\breve{G}([t]))|}{\tilde{h}_1([t], T)} \left(1 - \frac{\tilde{h}_1([t], T)}{\tilde{h}_1(t, T)} \right)$$

$$+ \sup_{\tilde{m} \leq t \leq \tilde{n}} \frac{|W(\breve{G}([t])) - W(\breve{G}(t))|}{\tilde{h}_1(t, T)}$$

$$= o_P(1),$$

where in the last line we took the following relations into account:

1. By the arguments of (2.4.24) we obtain

$$\sup_{\tilde{m} \leq t \leq \tilde{n}} \frac{|W(\breve{G}([t]))|}{\tilde{h}_1([t], T)} \overset{D}{=} \sup_{\tilde{m} \leq t \leq \tilde{n}} \frac{|W(\breve{G}([t])/\breve{G}(T/\mu_0))|}{(\breve{G}([t])/\breve{G}(T/\mu_0))^{\beta}} = \mathcal{O}_P(1).$$

2. Similar as in (2.4.6) it holds that

$$\sup_{\tilde{m} \leq t \leq \tilde{n}} \left(1 - \frac{\tilde{h}_1([t], T)}{\tilde{h}_1(t, T)} \right) \leq \frac{\beta}{\breve{G}([t])} \overset{\text{a.s.}}{=} o(1).$$

3. By the mean value theorem and Lemma 2.2.3 there is some $\xi \in \big[[t], t\big]$ such that

$$0 \leq \breve{G}(t) - \breve{G}([t]) \leq \breve{G}'(\xi) = \left(\frac{\int_0^\xi g(\xi) - g(x)\, dx}{\xi} \right)^2 \leq g(t)^2 \leq 1,$$

hence by Proposition 1.3.2 we find that

$$\sup_{\tilde{m} \leq t \leq \tilde{n}} \frac{|W(\breve{G}([t])) - W(\breve{G}(t))|}{\tilde{h}_1(t, T)} \overset{\text{a.s.}}{=} \frac{\sqrt{\log(T)}}{m^{\Lambda/2}} \, \mathcal{O}(1) = o(1).$$

The corresponding assertions for $\hat{\tau}_{m,-}^{closed}$ and $\hat{\tau}_m^{closed}$ follow by the same arguments.

□

Proof of Theorem 2.4.6. As in the proof of Theorem 2.4.5 it is sufficient to show the assertion for $\tilde{R}_k = (\hat{\sigma}_k/\sigma)\, \hat{R}_k$. Decomposing

$$\frac{\tilde{R}_k}{\hat{h}_2(k, \tilde{m})} = \frac{R_k}{\hat{h}_2(k, \tilde{m})} + \frac{\sum_{j=1}^{k} g(j)\, (\mu_0 - \hat{\mu}_{0,k})}{\sigma\, \hat{h}_2(k, \tilde{m})}$$

we aim to show that, for one thing,

$$\sup_{k=\tilde{m},...,\infty} \left| \frac{R_k}{\hat{h}_2(k,\tilde{m})} - \frac{\int_0^k g(x)\,dW(x)}{\hat{h}_2(k,\tilde{m})} \right| \stackrel{\text{a.s.}}{=} o(1) \qquad (2.4.27)$$

and that, for another thing,

$$\sup_{k=\tilde{m},...,\infty} \left| \frac{\sum_{j=1}^k g(j)\,(\mu_0 - \hat{\mu}_{0,k})}{\sigma\,\hat{h}_2(k,\tilde{m})} - \frac{\int_0^k g(x)dx\,W(k)/k}{\hat{h}_2(k,\tilde{m})} \right| \stackrel{\text{a.s.}}{=} o(1), \qquad (2.4.28)$$

which combines to

$$\sup_{k=\tilde{m},...,\infty} \left| \frac{\tilde{R}_k}{\hat{h}_2(k,\tilde{m})} - \frac{U(k)}{\hat{h}_2(k,\tilde{m})} \right| \stackrel{\text{a.s.}}{=} o(1),$$

where $U(t)$ is defined in (2.4.21). Analyzing the covariance structure of U (see (1.3.19)) we have $\{U(t)\,|\,t \geq \tilde{m}\} \stackrel{D}{=} \{W(\tilde{G}(t))\,|\,t \geq \tilde{m}\}$. Thus, if we show that

$$\sup_{t \geq \tilde{m}} \left| \frac{W(\tilde{G}([t]))}{\hat{h}_2([t],\tilde{m})} - \frac{W(\tilde{G}(t))}{\hat{h}_2(t,\tilde{m})} \right| = o_P(1) \qquad (2.4.29)$$

holds true, the assertion for $\tau_{m,+}^{open}$ follows by the arguments of (2.4.15) and the assertions for $\tau_{m,-}^{open}$ and τ_m^{open} follow in the same manner. In conclusion we find that the proof is complete if we show that (2.4.27), (2.4.28) and (2.4.29) hold true.

(2.4.27) follows along the lines of the proof of Theorem 2.4.2, yet we have to show that the convergence rates hold true for $\hat{h}_2(t,\tilde{m})$ instead of $h_2(t,\tilde{m})$. Concerning (2.4.11) we note that

$$\max_{k=\tilde{m},...,\infty} \frac{g(k)\,k^\kappa}{\hat{h}_2(k,\tilde{m})}\,\mathcal{O}(1) \stackrel{\text{a.s.}}{=} m^{\kappa-\Lambda/2}\,\mathcal{O}(1) = o(1).$$

As to (2.4.12) we note that by $\Lambda > 0$

$$\max_{k=\tilde{m},...,\infty} \frac{g(k)\,\sqrt{\log(k)}}{\hat{h}_2(k,\tilde{m})} \stackrel{\text{a.s.}}{=} \frac{\sqrt{\log(m)}}{m^{\Lambda/2}}\,\mathcal{O}(1) = o(1).$$

Furthermore, assertions (2.4.13), (2.4.14), (2.4.25) and (2.4.26) carry over to the case of $\hat{h}_2(t,\tilde{m})$ in the same manner, hence (2.4.28) and (2.4.29) hold true. $\qquad\square$

2.5   Consistency of the testing procedures

In this section we show that if the detectors contain a sufficient amount of per-turbed data, i.e. observations made after the change point, the testing procedures have asymptotic power 1. Sufficiently many perturbed observations means that $(\tilde{n} - k^*)/T$ needs to be a.s. bounded away from zero as $m \to \infty$.

In Section 2.5.1 and Section 2.5.2 we show consistency of the testing procedures for known and unknown in-control parameters. The assumptions in case of known in-control parameters are generally rather mild, whereas the assumptions on the slope parameters γ and λ in case of unknown in-control parameters are more restrictive.

2.5.1   Consistency for known in-control parameters

Theorem 2.5.1. *With the notation and assumptions of Section 2.1 and Section 2.2 and if, in addition,* $|\delta|\sqrt{T/\log\log(T)} \to \infty$ *and* $k^* \leq \theta T$ *for some* $0 < \theta < \min\{1/\mu_0, 1/\mu_1\}$, *it holds that*

$$under \ H_{1,+}^{closed}, \quad \lim_{m\to\infty} P\big(\tau_{m,+}^{closed} \leq \tilde{n}\big) = 1,$$

$$under \ H_{1,-}^{closed}, \quad \lim_{m\to\infty} P\big(\tau_{m,-}^{closed} \leq \tilde{n}\big) = 1,$$

$$under \ H_1^{closed}, \quad \lim_{m\to\infty} P\big(\tau_m^{closed} \leq \tilde{n}\big) = 1.$$

Hence, the tests are consistent.

Theorem 2.5.2. *With the notation and assumptions of Section 2.1 and Section 2.2 it holds that*

$$under \ H_{1,+}^{open}, \quad \lim_{m\to\infty} P\big(\tau_{m,+}^{open} < \infty\big) = 1,$$

$$under \ H_{1,-}^{open}, \quad \lim_{m\to\infty} P\big(\tau_{m,-}^{open} < \infty\big) = 1,$$

$$under \ H_1^{open}, \quad \lim_{m\to\infty} P\big(\tau_m^{open} < \infty\big) = 1.$$

Hence, the tests are consistent.

Proof of Theorem 2.5.1. Our assumption on k^* ensures that a.s. we have realiza-tions of $\tilde{S}_k$ that contain observations made after the change point: Without loss of generality, assume $\mu_0 < \mu_1$. As in Section 2.2.1 we denote by $N_{\mu_1}(t)$ the counting process based on $\{Y(n) + n\mu_1\}$. Then we have for sufficiently large T

$$\tilde{n} = T\frac{N(T)-1}{T} \geq T\frac{N_{\mu_1}(T)-1}{T} \overset{a.s.}{=} T\big(1/\mu_1 + o(1)\big) > T\theta = k^*. \qquad (2.5.1)$$

Hence, we can proceed under the assumption of $k^* < \tilde{n}$. We split the detector into two parts, where the first one only contains observations made before the change point and the second one also contains observations made after the change point:

$$\max_{k=\tilde{m},\dots,\tilde{n}} \frac{R_k}{h_1(k,T)} \geq \frac{R_{\tilde{n}}}{h_1(\tilde{n},T)} = \frac{R_{k^*}}{h_1(k^*,T)}\frac{h_1(k^*,T)}{h_1(\tilde{n},T)} + \frac{R_{\tilde{n}} - R_{k^*}}{h_1(\tilde{n},T)}.$$

The first summand is of order $\mathcal{O}_P(1)$, as $m \to \infty$. We take a closer look at the second summand: Using the decomposition of Remark 1.3.3 and the growth rates of (2.2.1) and Corollary 2.3.3 we obtain

$$
\begin{aligned}
|R_{\tilde{n}} - R_{k^*}| &\overset{\text{a.s.}}{=} \left| \sum_{i=k^*+1}^{\tilde{n}} \frac{g(i)\,(\mu(i) - \mu_0)}{\sigma} \right| + g(\tilde{n})\,\mathcal{O}\big(\sqrt{\tilde{n}}\,\log\log(\tilde{n})\big) \\
&\overset{\text{a.s.}}{\geq} \frac{|\mu(k^*+1) - \mu_0|}{\sigma} \sum_{i=k^*+1}^{\tilde{n}} g(i) + \mathcal{O}\big(\sqrt{T}\,\log\log(T)\big) \\
&\geq \frac{|\delta|\,(1 - 2^{-\gamma})}{\sigma} \sum_{i=k^*+1}^{\tilde{n}} g(i) + \mathcal{O}\big(\sqrt{T}\,\log\log(T)\big).
\end{aligned}
$$

Now, since $g(t)$ is increasing we have for sufficiently large T

$$
\sum_{i=k^*+1}^{\tilde{n}} g(i) \geq T\Big(\frac{\tilde{n}}{T} - \frac{k^*}{T}\Big)g(1) \overset{\text{a.s.}}{\geq} T(\min\{1/\mu_0, 1/\mu_1\} - \theta + o(1))\,g(1)
$$

and therefore

$$
\begin{aligned}
\max_{k=\tilde{m},\ldots,\tilde{n}} \frac{|R_k|}{h_1(k,T)} \\
\overset{\text{a.s.}}{\geq} \sqrt{T}\,|\delta|\,(d + o(1))\frac{\sqrt{T}}{h_1(\tilde{n},T)} + \mathcal{O}_P\big(\sqrt{T\log\log(T)}/h_1(\tilde{n},T)\big),
\end{aligned}
$$

where $d = (1 - 2^{-\gamma})\,g(1)\,(\min\{1/\mu_0, 1/\mu_1\} - \theta)/\sigma > 0$. Note that

$$
\frac{h_1(\tilde{n},T)}{\sqrt{T}} = \Big(\frac{G(\tilde{n})}{T}\Big)^{\beta} \Big(\frac{G(T/\mu_0)}{T}\Big)^{1/2-\beta}, \tag{2.5.2}
$$

where the second factor has a constant, positive limit and, by (2.2.1), the first factor is a.s. bounded below and above by positive constants for sufficiently large T. Combining this with $|\delta|\sqrt{T\log\log(T)} \to \infty$ completes the proof. $\qquad\square$

Proof of Theorem 2.5.2. Let $N = N_m$ be some sequence satisfying $k^* \leq \theta N$ for some $\theta < 1$ and $\delta\sqrt{N/\log(N)} \to \infty$, as $m \to \infty$. A possible choice for N is for example

$$
N = \begin{cases} k^*/(\theta\,|\delta_m|^3) & \text{if } \delta_m \to 0, \\ k^*/\theta & \text{if } \delta_m \not\to 0. \end{cases}
$$

As in the proof of Theorem 2.5.1 we have for sufficiently large N

$$
\begin{aligned}
|R_N| &\overset{\text{a.s.}}{\geq} \frac{|\delta|\,(1 - 2^{-\gamma})}{\sigma} \sum_{i=k^*+1}^{N} g(i) + \mathcal{O}_P\big(\sqrt{N\log\log(N)}\big) \\
&\geq |\delta|\,d\,(N - k^*) + \mathcal{O}_P\big(\sqrt{N\log\log(N)}\big),
\end{aligned}
$$

where $d = (1 - 2^{-\gamma}) g(1)/\sigma > 0$. Hence,

$$\frac{|R_N|}{h_2(N, \tilde{m})} \overset{\text{a.s.}}{\geq} |\delta| \, d \, \frac{N - k^*}{h_2(N, \tilde{m})} + \mathcal{O}_P\big(\sqrt{N \log \log(N)}/h_2(N, \tilde{m})\big)$$

$$\geq |\delta| \, d \, \theta \, \frac{N}{h_2(N, \tilde{m})} + \mathcal{O}_P\big(\sqrt{N \log \log(N)}/h_2(N, \tilde{m})\big).$$

Finally, the assertions follow by $h_2(N, \tilde{m}) \overset{\text{a.s.}}{\simeq} \sqrt{N \log(N/m)}$ (cf. Remark 1.7.1) and our assumptions on N . $\square$

2.5.2 Consistency for unknown in-control parameters

Theorem 2.5.3. *With the notation and assumptions of Section 2.1 and Section 2.2 and, in addition,* $k^* \leq \theta\,T$ *for some* $0 < \theta < \min\{1/\mu_0, 1/\mu_1\}$,

$$\frac{|\delta|\,T^{1/2-\lambda-\gamma}}{\sqrt{\log\log(T)}} \to \infty,$$

as $m \to \infty$, *and* $\hat{\sigma}_k$ *being an estimate such that*

$$\frac{|\delta|\,T^{1/2-\gamma}}{\hat{\sigma}_{\tilde{n}}} \xrightarrow{P} \infty,$$

as $m \to \infty$, *we have*

$$\text{under } H_{1,+}^{closed}, \quad \lim_{m\to\infty} P\big(\hat{\tau}_{m,+}^{closed} \leq \tilde{n}\big) = 1,$$
$$\text{under } H_{1,-}^{closed}, \quad \lim_{m\to\infty} P\big(\hat{\tau}_{m,-}^{closed} \leq \tilde{n}\big) = 1,$$
$$\text{under } H_{1}^{closed}, \quad \lim_{m\to\infty} P\big(\hat{\tau}_{m}^{closed} \leq \tilde{n}\big) = 1.$$

Hence, the tests are consistent.

Theorem 2.5.4. *With the notation and assumptions of Section 2.1 and Section 2.2 and, in addition, some sequence* $N = N_m$ *satisfying* $k^* \leq \theta\,N$ *for some* $\theta < 1$,

$$\frac{|\delta|\,N^{1/2-\lambda-\gamma}}{\sqrt{\log\log(N)}} \to \infty,$$

as $m \to \infty$, *and* $\hat{\sigma}_k$ *being an estimate such that*

$$\frac{|\delta|\,N^{1/2-\gamma}}{\sqrt{\log(N/m)}\,\hat{\sigma}_N} \xrightarrow{P} \infty,$$

as $m \to \infty$, *we have*

$$\text{under } H_{1,+}^{open}, \quad \lim_{m\to\infty} P\big(\hat{\tau}_{m,+}^{open} < \infty\big) = 1,$$
$$\text{under } H_{1,-}^{open}, \quad \lim_{m\to\infty} P\big(\hat{\tau}_{m,-}^{open} < \infty\big) = 1,$$
$$\text{under } H_{1}^{open}, \quad \lim_{m\to\infty} P\big(\hat{\tau}_{m}^{open} < \infty\big) = 1.$$

Hence, the tests are consistent.

Remark 2.5.5. *If* $\hat{\sigma}_k = \hat{\sigma}_{\tilde{m}}$ *for all* k, *i.e. if we estimate* σ *only by observations made in the training period, and if* $\lambda + \gamma < 1/2$, *a possible choice for* N *in Theorem 2.5.4 is*

$$N = \begin{cases} k^*/(\theta\,|\delta_m|^{1/(1-\lambda-\gamma)}) & \text{if } \delta_m \to 0, \\ k^*/\theta & \text{if } \delta_m \not\to 0. \end{cases}$$

Proof of Theorem 2.5.3. The first part of the proof follows along the lines of the proof of Theorem 2.5.1, i.e. we have $k^* < \tilde{n}$ a.s. We focus on the detectors with known variance, i.e. $\tilde{R}_k = (\hat{\sigma}_k/\sigma)\,\hat{R}_k$, first and replace σ by its estimate later on in the proof. Via the decomposition of Remark 1.3.3, the growth rates of (2.2.1) and Corollary 2.3.3 and Theorem 2.4.1 we obtain for $k^* < \tilde{n}$

$$
\begin{aligned}
\tilde{R}_{\tilde{n}} &= \tilde{R}_{k^*} + R_{\tilde{n}} - R_{k^*} \\
&\quad + \sum_{i=1}^{k^*} \frac{g(i)\,(\tilde{S}_{k^*}/k^* - \mu_0)}{\sigma} - \sum_{i=1}^{\tilde{n}} \frac{g(i)\,(\tilde{S}_{\tilde{n}}/\tilde{n} - \mu_0)}{\sigma} \\
&= \mathcal{O}_P\big(\hat{h}_1(k^*,T)\big) + \sum_{i=1}^{\tilde{n}} \frac{g(i)(\mu(i)-\mu_0)}{\sigma} \\
&\quad + \mathcal{O}_P\big(\sqrt{T\log\log(T)}\big) - \sum_{i=1}^{\tilde{n}} \frac{g(i)(\sum_{j=1}^{\tilde{n}}\mu(j)/\tilde{n} - \mu_0)}{\sigma} \\
&= \hat{\Lambda}_{\tilde{n}} + \mathcal{O}_P\big(\sqrt{T\log\log(T)}\big),
\end{aligned}
\tag{2.5.3}
$$

where

$$
\hat{\Lambda}_k := \sum_{i=1}^{k} \frac{g(i)\left(\mu(i) - \sum_{j=1}^{k}\mu(j)/k\right)}{\sigma}.
$$

Hence, we have to find the rate of convergence of $\hat{\Lambda}_{\tilde{n}}$. To shorten the notation we introduce

$$
\bar{\mu}_k = \frac{1}{k}\sum_{i=1}^{k}\mu(i) \quad \text{and} \quad \bar{g}_k = \frac{1}{k}\sum_{i=1}^{k}g(i).
$$

By elementary calculations we obtain the following representation of the increments of $\hat{\Lambda}_k$ for all $k = 1,\dots,\tilde{n}-1$:

$$
\begin{aligned}
\hat{\Lambda}_{k+1} - \hat{\Lambda}_k &= \sum_{i=1}^{k+1} \frac{g(i)\,(\mu(i) - \bar{\mu}_{k+1})}{\sigma} - \sum_{i=1}^{k} \frac{g(i)\,(\mu(i) - \bar{\mu}_k)}{\sigma} \\
&= \frac{k}{k+1}\,\frac{(g(k+1) - \bar{g}_k)\,(\mu(k+1) - \bar{\mu}_k)}{\sigma}.
\end{aligned}
\tag{2.5.4}
$$

Before the change point the mean function $\mu(k)$ is constant, thus $\mu(k) - \bar{\mu}_k = 0$ for $k \leq k^*$ and therefore $\hat{\Lambda}_k = 0$ for $k \leq k^*$. Hence, we can express $\hat{\Lambda}_{\tilde{n}}$ as the following telescoping series and make use of representation (2.5.4):

$$
\hat{\Lambda}_{\tilde{n}} = \sum_{i=k^*}^{\tilde{n}} \hat{\Lambda}_i - \hat{\Lambda}_{i-1} = \sum_{i=k^*}^{\tilde{n}} \frac{i}{i+1}\,\frac{(g(i+1) - \bar{g}_i)\,(\mu(i+1) - \bar{\mu}_i)}{\sigma}.
$$

For the sake of simplicity, we restrict ourselves to the case of $\delta > 0$, i.e. the case of $\mu(k)$ being increasing. The weight function is assumed to be increasing, too, so it holds that $\mu(i+1) - \bar{\mu}_i \geq 0$ and $g(i+1) - \bar{g}_i \geq 0$, hence

$$\hat{\Lambda}_{\tilde{n}} \geq \frac{1}{2\sigma} \sum_{i=k^*}^{\tilde{n}} \left(g(i+1) - \bar{g}_i\right)\left(\mu(i+1) - \bar{\mu}_i\right).$$

We take a closer look at the two factors separately: On the one hand, we have

$$
\begin{aligned}
& g(i+1) - \bar{g}_i \\
&= -(2+i)^{-\lambda} + \frac{1}{i}\sum_{j=1}^{i}(1+j)^{-\lambda} \\
&\geq -(2+i)^{-\lambda} + \frac{1}{i}\int_1^{i+1}(1+x)^{-\lambda}\,dx \\
&= -(2+i)^{-\lambda} + \frac{(2+i)^{1-\lambda} - 2^{1-\lambda}}{(1-\lambda)\,i} \\
&= (2+i)^{-\lambda}\left(\frac{1}{1-\lambda}-1\right) + \frac{(2+i)^{-\lambda}}{1-\lambda}\left(\frac{2+i}{i}-1\right) - \frac{2^{1-\lambda}}{(1-\lambda)\,i} \\
&= \frac{\lambda}{1-\lambda}(2+i)^{-\lambda} + \frac{(2+i)^{-\lambda}}{1-\lambda}\frac{2}{i} - \frac{2^{1-\lambda}}{(1-\lambda)\,i}.
\end{aligned}
$$

On the other hand, it holds that

$$
\begin{aligned}
& \mu(i+1) - \bar{\mu}_i \\
&= \delta\left(-(2+i-k^*)^{-\gamma} + \frac{1}{i}\left(k^* + \sum_{j=1}^{i-k^*}(1+j)^{-\gamma}\right)\right) \\
&\geq \delta\left(-(2+i-k^*)^{-\gamma} + \frac{1}{i}\left(k^* + \int_1^{i-k^*+1}(1+x)^{-\gamma}\,dx\right)\right) \\
&= \delta\left(-(2+i-k^*)^{-\gamma} + \frac{1}{i}\left(k^* + \frac{(2+i-k^*)^{1-\gamma} - 2^{1-\gamma}}{1-\gamma}\right)\right) \qquad (2.5.5) \\
&= \delta\,(2+i-k^*)^{-\gamma}\left(\frac{1}{1-\gamma}-1\right) \\
&\quad + \delta\left(\frac{(2+i-k^*)^{-\gamma}}{1-\gamma}\left(\frac{2+i-k^*}{i}-1\right) + \frac{k^*}{i} - \frac{2^{1-\gamma}}{(1-\gamma)\,i}\right) \\
&\geq \frac{\delta\,\gamma}{1-\gamma}(2+i)^{-\gamma} + \delta\left(\frac{k^*}{i}\left(1 - \frac{(2+i-k^*)^{-\gamma}}{1-\gamma}\right) - \frac{2^{1-\gamma}}{(1-\gamma)\,i}\right).
\end{aligned}
$$

The second sum of the right hand side of (2.5.5) is positive (hence can be dropped) if and only if $i > k^* + \Gamma$, where $\Gamma := (1-\gamma)^{-1/\gamma} - 2$ is constant. So, for one thing, we have

$$\left|\sum_{i=k^*}^{k^*+\lceil\Gamma\rceil}\left(g(i+1)-\bar{g}_i\right)\left(\mu(i+1)-\bar{\mu}_i\right)\right| \leq (\Gamma+1)\,|\mu_0-\mu_1| = \mathcal{O}(1),$$

and, for another thing, we have

$$\sum_{i=k^*+[\Gamma]+1}^{\tilde{n}} (g(i+1) - \bar{g}_i) (\mu(i+1) - \bar{\mu}_i)$$

$$\geq \sum_{i=k^*+[\Gamma]+1}^{\tilde{n}} \frac{\lambda \gamma \delta}{(1-\lambda)(1-\gamma)} (2+i)^{-(\lambda+\gamma)} + r(i), \tag{2.5.6}$$

where, for some constant $d > 0$, $|r(i)| \leq d \, i^{-\min\{\lambda,\gamma\}-1}$ and therefore

$$\left| \sum_{i=k^*+[\Gamma]+1}^{\tilde{n}} r(i) \right| \leq d \int_{k^*}^{\tilde{n}+1} x^{-\min\{\lambda,\gamma\}-1} \, dx$$

$$= d \, \frac{k^{*-\min\{\lambda,\gamma\}} - \tilde{n}^{-\min\{\lambda,\gamma\}}}{\min\{\lambda,\gamma\}}$$

$$= \frac{d}{\min\{\lambda,\gamma\}} k^{*-\min\{\lambda,\gamma\}} \left(1 - (k^*/\tilde{n})^{-\min\{\lambda,\gamma\}}\right)$$

$$\overset{\text{a.s.}}{=} \mathcal{O}\left(m^{-\min\{\lambda,\gamma\}}\right)$$

$$= o(1).$$

The first summand on the right hand side of (2.5.6) gives us the required rate of convergence:

$$\sum_{i=k^*+[\Gamma]+1}^{\tilde{n}} (2+i)^{-(\lambda+\gamma)}$$

$$\geq \int_{k^*+[\Gamma]+1}^{\tilde{n}-2} (2+x)^{-(\lambda+\gamma)} \, dx$$

$$= \frac{\tilde{n}^{1-(\lambda+\gamma)} - (k^*+[\Gamma]+3)^{1-(\lambda+\gamma)}}{1-(\lambda+\gamma)}$$

$$\geq \frac{\tilde{n}^{1-(\lambda+\gamma)} - (\theta T + [\Gamma]+3)^{1-(\lambda+\gamma)}}{1-(\lambda+\gamma)}$$

$$= \frac{T^{1-(\lambda+\gamma)}}{1-(\lambda+\gamma)} \left(\left(\frac{\tilde{n}}{T}\right)^{1-(\lambda+\gamma)} - \left(\frac{\theta T + [\Gamma]+3}{T}\right)^{1-(\lambda+\gamma)} \right)$$

$$\overset{\text{a.s.}}{\geq} \frac{T^{1-(\lambda+\gamma)}}{1-(\lambda+\gamma)} \left((1/\mu_0)^{1-(\lambda+\gamma)} - \theta^{1-(\lambda+\gamma)} + o(1) \right),$$

where by our assumptions on θ it holds true that $(1/\mu_0)^{1-(\lambda+\gamma)} - \theta^{1-(\lambda+\gamma)} > 0$. Combining the inequalities above yields

$$\hat{\Lambda}_{\tilde{n}} \overset{\text{a.s.}}{\geq} (\tilde{d} + o(1)) \, \delta \, T^{1-\lambda-\gamma}, \tag{2.5.7}$$

where

$$\tilde{d} = \frac{\lambda \gamma \left((1/\mu_0)^{1-(\lambda+\gamma)} - \theta^{1-\lambda-\gamma} \right)}{2\sigma(1-\lambda)(1-\gamma)(1-\lambda-\gamma)} > 0.$$

Plugging this relation into (2.5.3) yields

$$
\frac{|\hat{R}_{\tilde{n}}|}{\hat{h}_1(\tilde{n}, T)}
$$
$$
\overset{\text{a.s.}}{\geq} (\check{d} + o(1)) \frac{\delta\, T^{1-\lambda-\gamma}}{\hat{h}_1(\tilde{n}, T)\, \hat{\sigma}_{\tilde{n}}} + \mathcal{O}_P\big(\sqrt{T \log\log(T)}/(\hat{h}_1(\tilde{n}, T)\, \hat{\sigma}_{\tilde{n}})\big).
$$

By the definition of $\tilde{n}$ (see Section 2.2.1) we have

$$
\tilde{S}_{\tilde{n}} = \inf\{t \in \mathcal{I}_c \,|\, N(t) > N(T) - 1\} = T,
$$

thus the estimate $\hat{\mu}_{0,\tilde{n}}$ and the threshold function simplify to

$$
\hat{\mu}_{0,\tilde{n}} = \tilde{S}_{\tilde{n}}/\tilde{n} = T/\tilde{n}
$$

and

$$
\hat{h}_1(\tilde{n}, T) = \tilde{G}^{\beta}(\tilde{n})\big(\tilde{G}(T/\hat{\mu}_{0,\tilde{n}})\big)^{1/2-\beta} = \tilde{G}^{1/2}(\tilde{n}) \overset{\text{a.s.}}{\simeq} \tilde{n}^{1/2-\lambda}.
$$

Hence, by (2.2.1) it holds for some $\check{d} > 0$,

$$
\frac{\hat{R}_{\tilde{n}}}{\hat{h}_1(\tilde{n}, T)} \overset{\text{a.s.}}{\geq} (\check{d} + o(1)) \frac{\delta\, T^{1/2-\gamma}}{\hat{\sigma}_{\tilde{n}}} + \mathcal{O}_P(\sqrt{\log\log(T)}\, T^{\lambda}/\hat{\sigma}_{\tilde{n}}).
$$

By the first assumption on the convergence of T in Theorem 2.5.3 we know that the deterministic part on the right hand side of the upper expression dominates the stochastic part and by the second assumption on the convergence of T we know that the deterministic part converges to infinity, hence (if $\delta > 0$)

$$
\frac{\hat{R}_{\tilde{n}}}{\hat{h}_1(\tilde{n}, T)} \overset{P}{\to} \infty.
$$

The case of $\delta < 0$ follows in the same manner. $\qquad\square$

Proof of Theorem 2.5.4. As in the proof of Theorem 2.5.1 we have

$$
\begin{aligned}
|\hat{R}_N| &\overset{\text{a.s.}}{\geq} (d + o(1))\frac{|\delta|\, N^{1-\lambda-\gamma}}{\hat{h}_2(N, \tilde{m})\, \hat{\sigma}_N} \\
&\quad + \mathcal{O}_P\big(\sqrt{N \log\log(N)}/(\hat{h}_2(N, \tilde{m})\, \hat{\sigma}_N)\big) \\
&= (\check{d} + o(1))\frac{|\delta|\, N^{1/2-\gamma}}{\sqrt{\log(N/m)}\, \hat{\sigma}_N} \\
&\quad + \mathcal{O}_P\big(\sqrt{\log\log(N)}\, T^{\lambda}/\big(\sqrt{\log(N/m)}\, \hat{\sigma}_N\big)\big)
\end{aligned}
$$

for some $\tilde{d}, \check{d} > 0$, which, by our assumptions on N yields the assertion. $\qquad\square$

2.6 Asymptotic normality of the delay times

In this section we restrict ourselves to an early-change setting (see (2.6.2) and (2.6.3)), in which we are able show that the delay times of the testing procedures are asymptotically normal. The early-change assumption in the closed-end setting is somewhat stronger because the interplay between T (the total number of observations) and k^* (the change point) has to be taken into account.

So far, the results in the closed-end setting are based on a truncation point $T = [\vartheta\, m^\rho]$ where $\rho \geq 1$ (see Section 2.1), however, for the (closed-end) results below we have to tighten this condition to $\rho > 1$. The reason for this restriction can be found e.g. in the arguments of Lemma 2.6.14, where we prove that (in the early change setting) the probability to stop before the change point is negligible.

For the sake of readability, the results of Section 2.6.1 and Section 2.6.2 are given at the beginning of the respective section, whereas all the proofs are postponed to Section 2.6.1.1 or Section 2.6.2.1, respectively.

On standardizing the delay times we introduce the function

$$r : [1, \infty) \to [-\gamma/(1-\gamma), \infty), \quad x \mapsto x - \frac{x^{1-\gamma}}{1-\gamma}, \tag{2.6.1}$$

which is discussed further in Lemma 2.6.6 below.

2.6.1 Asymptotic normality of the delay times for known in-control parameters

Theorem 2.6.1. *With the notation and assumptions of Section 2.1 and Section 2.2 as well as* $T = [\vartheta\, m^\rho]$ *where* $1 < \rho < (3/2 - 2\beta)/(1 - 2\beta)$ *and* $\vartheta \geq 1$, δ *being constant and being estimated by* $\hat{\delta} = \delta + o_P\big(m^{(1-\rho)(1/2-\beta)}\big)$, *the critical value* c *being positive and*

$$k^* = m/\mu_0 + \mathcal{O}\big(m^\eta\big) \tag{2.6.2}$$

for some $0 < \eta < 1 - (\rho - 1)(1/2 - \beta)$ *it holds for all* $x \in \mathbb{R}$ *under the alternative* $H_{1,+}^{closed}$ *that*

$$\lim_{m \to \infty} P\left(\frac{\hat{\delta}\, g(\tilde{m})\, r\big(\max\{\tau_{m,+}^{closed} - k^*, 1\}\big)/\sigma - c\, h_1(\tilde{m}, T)}{\sqrt{G(\tilde{m})}} \leq x \right) = \phi(x)$$

and under the alternative $H_{1,-}^{closed}$ *that*

$$\lim_{m \to \infty} P\left(\frac{|\hat{\delta}|\, g(\tilde{m})\, r\big(\max\{\tau_{m,-}^{closed} - k^*, 1\}\big)/\sigma - c\, h_1(\tilde{m}, T)}{\sqrt{G(\tilde{m})}} \leq x \right) = \phi(x),$$

where $r(x)$ *is defined in* (2.6.1).

Theorem 2.6.2. *With the notation and assumptions of Section 2.1 and Section 2.2 as well as δ being constant and being estimated by $\hat{\delta} \xrightarrow{P} \delta$, the parameter c in the threshold function h_2 being positive and*

$$k^* = m/\mu_0 + \mathcal{O}(m^\eta) \tag{2.6.3}$$

for some $\eta < 1$ it holds for all $x > -c$ under the alternative $H_{1,+}^{open}$ that

$$\lim_{m\to\infty} P\left(\frac{\hat{\delta}\, g(\tilde{m})\, r\big(\max\{\tau_{m,+}^{open} - k^*, 1\}\big)/\sigma - c\sqrt{G(\tilde{m})}}{\sqrt{G(\tilde{m})}} \leq x \right) = \phi(x)$$

and under the alternative $H_{1,-}^{open}$ that

$$\lim_{m\to\infty} P\left(\frac{|\hat{\delta}|\, g(\tilde{m})\, r\big(\max\{\tau_{m,-}^{open} - k^*, 1\}\big)/\sigma - c\sqrt{G(\tilde{m})}}{\sqrt{G(\tilde{m})}} \leq x \right) = \phi(x),$$

where $r(x)$ is defined in (2.6.1).

In the following lemma we suggest a suitable estimate which fulfills the assumptions of Theorem 2.6.1 or Theorem 2.6.2, respectively.

Lemma 2.6.3. *With the notation and assumptions of Section 2.1 and Section 2.2 let*

$$\hat{\delta}_{m,k} = \frac{\sum_{i=1}^{k} \mu_{\tilde{m}}(i)\left(\tilde{S}_i - \tilde{S}_{i-1} - \mu_0 \right)}{\sum_{i=1}^{k} \mu_{\tilde{m}}^2(i)}, \tag{2.6.4}$$

where

$$\mu_l(k) := 1 - (1 + (k-l)_+)^{-\gamma} = \begin{cases} 1 - (1 + k - l)^{-\gamma} & \text{if } k > l, \\ 0 & \text{if } k \leq l. \end{cases}$$

If $k^* - m = o(m^\zeta)$, as $m \to \infty$, for some $\zeta > 1$ it holds that, as $m \to \infty$,

$$\hat{\delta}_{m,\tilde{m}+m^\zeta} \overset{a.s.}{=} \delta + \mathcal{O}(m^{1-\zeta}) + \mathcal{O}(m^{-\zeta\gamma}) + \mathcal{O}(\sqrt{\log\log(m)}\, m^{-\zeta/2}).$$

On choosing

$$\zeta > \max\left\{ 1 + (\rho-1)(1/2 - \beta), (\rho-1)(1/2 - \beta)/\gamma, (\rho-1)(1 - 2\beta) \right\}$$

the assumptions on $\hat{\delta}$ of Theorem 2.6.1 are fulfilled and on choosing $\zeta > 1$ the assumptions of Theorem 2.6.2 are fulfilled.

Since a slightly more general version of Lemma 2.6.3 (i.e. with unknown μ_0) is shown in Section 2.6.2 the proof is omitted here. Combining Lemma 2.6.3 with Theorem 2.6.1 or Theorem 2.6.2, respectively, gives us the following confidence intervals for the change point:

Corollary 2.6.4. *Let the assumptions of Theorem 2.6.1 hold true. On setting*

$$q = r^{-1}\left(\frac{\sqrt{G(\tilde{m})}\, \phi^{-1}(1-\tilde{\alpha}) + c\, h_1(\tilde{m}, T)}{|\hat{\delta}|\, g(\tilde{m})/\sigma} \right),$$

where $\tilde{\alpha} \in (0,1)$, *we obtain for* $\tau_{m,+}^{closed}$, $\tau_{m,-}^{closed}$ *and* τ_m^{closed} *as in Section 2.2.2 that, as* $m \to \infty$,

$$P\big(\tau_{m,+}^{closed} - q \leq k^* < \tau_{m,+}^{closed}\big) = 1 - \tilde{\alpha} + o(1),$$
$$P\big(\tau_{m,-}^{closed} - q \leq k^* < \tau_{m,-}^{closed}\big) = 1 - \tilde{\alpha} + o(1),$$
$$P\big(\tau_m^{closed} - q \leq k^* < \tau_m^{closed}\big) \geq 1 - \tilde{\alpha} + o(1),$$

hold true under $H_{1,+}^{closed}$, $H_{1,-}^{closed}$ *or* H_1^{closed}, *respectively.*

Corollary 2.6.5. *Let the assumptions of Theorem 1.5.1 hold true. On setting*

$$q(\chi_i) = r^{-1}\left(\frac{\sqrt{G(\tilde{m})}\, \phi^{-1}(\chi_i) + G(\tilde{m})}{|\hat{\delta}|\, g(\tilde{m})/\sigma} \right),$$

where

$$\chi_i = \begin{cases} \phi^{-1}\big(1 - \tilde{\alpha} - \phi(-c)\big) & \text{if } i = 1, \\ \phi^{-1}\big(1 - \tilde{\alpha} - 2\phi(-c)\big) & \text{if } i = 2, \end{cases}$$

with

$$\tilde{\alpha} \in \begin{cases} (\phi(-c),\, 1) & \text{for } i = 1, \\ (2\,\phi(-c),\, 1) & \text{for } i = 2, \end{cases} \tag{2.6.5}$$

we obtain for $\tau_{m,+}^{open}$ *and* $\tau_{m,-}^{open}$ *as in Section 2.2.2 and* $\tilde{\tau}_m^{open} := \min\{\tau_{m,+}^{open}, \tau_{m,-}^{open}\}$ *that, as* $m \to \infty$,

$$P\big(\tau_{m,+}^{open} - q(\chi_1) \leq T^* < \tau_{m,+}^{open}\big) = 1 - \tilde{\alpha} + o(1),$$
$$P\big(\tau_{m,-}^{open} - q(\chi_1) \leq T^* < \tau_{m,+}^{open}\big) = 1 - \tilde{\alpha} + o(1),$$
$$P\big(\tilde{\tau}_m^{open} - q(\chi_2) \leq T^* < \tilde{\tau}_m^{open}\big) \geq 1 - \tilde{\alpha} + o(1),$$

hold true under $H_{1,+}^{open}$, $H_{1,-}^{open}$ *or* H_1^{open}, *respectively.*

Note that in this section we assume σ to be known. Confidence intervals where σ is being estimated as well can be found in Section 2.6.2.

2.6.1.1   Proofs

Before we turn to the proofs of the two main theorems, we introduce some notation which we will use throughout this section. Let

$$H := k^* + \big[w(k^*)\big] \tag{2.6.6}$$

with

$$w(k^*) := r^{-1}\left(\frac{\sqrt{G(k^*)}\,\sigma}{g(k^*)\,|\delta|}\left(\frac{h(k^*)}{\sqrt{G(k^*)}} + x\right)\right), \tag{2.6.7}$$

where $h(t)$ shall denote either $c\,h_1(t,T)$ or $h_2(t,\tilde{m})$, depending on the setting we are working with. If not stated otherwise, throughout this section w shall denote $w(k^*)$. The idea of the proofs of Theorem 2.6.1 and Theorem 2.6.2 is to show $P(\tau \leq H) \to \phi(x)$ and then solve the inequality $\tau \leq H(x)$ for x (where τ shall indicate the respective stopping time). Remark 2.6.7, below, ensures that the expression w in (2.6.7) is well-defined.

To further shorten the notation, let

$$\Lambda_k := \sum_{i=1}^{k} \frac{g(i)\,(\mu(i) - \mu_0)}{\sigma} \tag{2.6.8}$$

be the deterministic perturbation of the detectors and

$$A(a,b) := \left(\max_{k=[a],\dots,[b]} \frac{\operatorname{sgn}(\delta)\,R_k}{h(k)} - \frac{\operatorname{sgn}(\delta)\,\Lambda_H}{h(H)}\right)\frac{h(k^*)}{\sqrt{G(k^*)}} \tag{2.6.9}$$

be the difference between the test statistic and the perturbation at time point H, where again the expression $h(t)$ shall denote either $c\,h_1(t,T)$ or $h_2(t,\tilde{m})$ depending on the setting we are working with.

The following lemma will help us handling the term w (defined in (2.6.7)).

Lemma 2.6.6. *The function*

$$r : [1,\infty) \to [-\gamma/(1-\gamma),\infty),\ x \mapsto x - x^{1-\gamma}/(1-\gamma)$$

has an increasing, continuous inverse function $r^{-1} : [-\gamma/(1-\gamma),\infty) \to [1,\infty)$ for which $r^{-1}(x) \sim x$ holds true as $x \to \infty$.

Proof. The existence of r^{-1} follows immediately by r being continuous and increasing, where the latter can be seen via the derivative r'. As to the asymptotic behavior of r^{-1}, we note that, on the one hand,

$$x = r\big(r^{-1}(x)\big) = r^{-1}(x) - \frac{\big(r^{-1}(x)\big)^{1-\gamma}}{1-\gamma} < r^{-1}(x),$$

hence

$$1 < \frac{r^{-1}(x)}{x}.$$

On the other hand, for $\xi \in (1-\gamma, 1)$ and x sufficiently large it holds that

$$r\big(x + x^\xi\big) = x + x^\xi - \frac{\big(x + x^\xi\big)^{1-\gamma}}{1-\gamma} > x,$$

which yields for x sufficiently large that

$$\frac{r^{-1}(x)}{x} < \frac{x + x^{\xi}}{x} \to 1$$

as $x \to \infty$. $\qquad\qquad\qquad\qquad\qquad\qquad\qquad\qquad\qquad\qquad\qquad\qquad$ $\square$

Remark 2.6.7. *By Lemma 2.6.6 we see that $w(k^*)$ is well-defined if*

$$\frac{\sqrt{G(k^*)}\,\sigma}{g(k^*)\,|\delta|}\left(\frac{h(k^*)}{\sqrt{G(k^*)}} + x\right) \geq -\frac{\gamma}{1-\gamma},$$

which is equivalent to

$$x \geq -\frac{\gamma\,g(k^*)\,|\delta|}{(1-\gamma)\sqrt{G(k^*)}\,\sigma} - \frac{h(k^*)}{\sqrt{G(k^*)}}.$$

On the one hand, we have $g(k^*)/\sqrt{G(k^*)} \to 0$, and, on the other hand, we have either

$$\frac{c\,h_1(k^*,T)}{\sqrt{G(k^*)}} = c\left(\frac{G(T/\mu_0)}{G(k^*)}\right)^{1/2-\beta} \to \infty$$

or

$$\frac{h_2(k^*,\tilde{m})}{\sqrt{G(k^*)}} = f^{-1}(\ln(G(k^*)/G(\tilde{m})) + f(c)) \overset{a.s.}{\geq} c$$

so in the closed-end setting, if the critical value is positive, the expression w is well-defined for any x, yet, in the open-end setting, one has to restrict the considerations to the case where $x > -c$.

The proofs of Theorem 2.6.1 and Theorem 2.6.2 are each subdivided into a sequence of lemmata given in the two paragraphs below.

Proof of Theorem 2.6.1

We begin with some technical but important details on the interplay of the convergence rates of the sequences m, k^*, w and H. Note that H is chosen such that (2.6.13) is fulfilled.

Lemma 2.6.8. *Under the assumptions of Theorem 2.6.1 we have, as $m \to \infty$,*

$$w \simeq h_1(k,T) \simeq (G(k^*))^{\beta}\,(G(T/\mu_0))^{\frac{1}{2}-\beta} \to \infty, \qquad\qquad (2.6.10)$$

$$H \sim k^* \sim m/\mu_0, \qquad\qquad\qquad\qquad\qquad\qquad\qquad\qquad (2.6.11)$$

$$h_1(H,T)/h_1(k^*,T) = \big(G(H)/G(k^*)\big)^{\beta} = 1 + \mathcal{O}(w/m), \qquad (2.6.12)$$

$$\left(\operatorname{sgn}(\delta)\,c - \frac{\Lambda_H}{h_1(H,T)}\right)\frac{h_1(k^*,T)}{\sqrt{G(k^*)}} \to -\operatorname{sgn}(\delta)\,x, \qquad (2.6.13)$$

where H, w and Λ_k are defined in (2.6.6), (2.6.7) and (2.6.8).

Proof of Lemma 2.6.8. Recall that

$$w = r^{-1}\left(\frac{\sqrt{G(k^*)}\,\sigma}{g(k^*)\,|\delta|} \left(\frac{c\,h_1(k^*)}{\sqrt{G(k^*)}} + x \right) \right).$$

Via Lemma 2.2.3 and the early-change assumption (2.6.2) we have, as $m \to \infty$,

$$\frac{h_1(k^*, T)}{\sqrt{G(k^*)}} = \left(\frac{G(T/\mu_0)}{G(k^*)} \right)^{1/2-\beta} \to \infty.$$

Combining this with Lemma 2.6.6 yields, as $m \to \infty$,

$$w \sim \frac{\sqrt{G(k^*)}\,\sigma}{|\delta|\,g(k^*)} \left(\frac{h_1(k^*, T)}{\sqrt{G(k^*)}} + x \right) \simeq h_1(k^*, T),$$

which shows the first assertion. Combining this with the early-change assumption, our assumption on ρ and Lemma 2.2.3 we have

$$0 \le \frac{w}{k^*} = \frac{k^{*\beta}\,T^{1/2-\beta}}{k^*}\,\mathcal{O}(1) = m^{\beta-1+\rho(1/2-\beta)}\,\mathcal{O}(1) = o(1),$$

which yields the second assertion. As to the third one, we first note that for $\beta = 0$ it holds true that $h_1(H,T)/h_1(k^*,T) = 1$. If $0 < \beta < 1/2$, on the other hand, it holds by the mean value theorem that some $\xi \in [k^*, H]$

$$\begin{aligned}
\frac{h_1(H,T)}{h_1(k^*,T)} - 1 &= \left(\frac{G(H)}{G(k^*)} \right)^{\beta} - 1 \\
&= \frac{G^{\beta}(H) - G^{\beta}(k^*)}{H - k^*}\,\frac{H - k^*}{G^{\beta}(k^*)} \\
&= \beta\,G'(\xi)\,G^{\beta-1}(\xi)\,\frac{w}{G^{\beta}(k^*)} \\
&= \mathcal{O}(w/m).
\end{aligned} \qquad (2.6.14)$$

The last assertion of Lemma 2.6.8 requires some more effort: Since $\mu(k) = \mu_0$ for $k \le k^*$, we have

$$\begin{aligned}
\Lambda_H &= \sum_{i=k^*+1}^{H} \frac{g(i)(\mu(i) - \mu_0)}{\sigma} \\
&= \frac{g(k^*)}{\sigma} \sum_{i=k^*+1}^{H} (\mu(i) - \mu_0) + \mathcal{O}\big((g(H) - g(k^*))\,w\big) \\
&= \frac{\delta\,g(k^*)}{\sigma} \sum_{i=1}^{[w]} \big(1 - (1+i)^{-\gamma}\big) + \mathcal{O}\big(g'(m)w^2\big) \\
&= \frac{\delta\,g(k^*)}{\sigma} \left([w] - \frac{[w]^{1-\gamma}}{1-\gamma} + \mathcal{O}\big(w^{-\gamma}\big) \right) + \mathcal{O}\big(m^{-\lambda-1}w^2\big) \\
&= \frac{\delta\,g(k^*)\,r([w])}{\sigma} + \mathcal{O}\big(w^{-\gamma} + m^{-\lambda-1}w^2\big).
\end{aligned}$$

Now, an application of the mean value theorem gives us that $r([w]) = r(w) + \mathcal{O}(1)$, hence

$$\Lambda_H = \frac{\delta\, g(k^*)\, r(w)}{\sigma} + \mathcal{O}\big(w^{-\gamma} + m^{-\lambda-1}w^2 + 1\big). \tag{2.6.15}$$

Plugging this relation into (2.6.13) yields

$$
\begin{aligned}
\left(\mathrm{sgn}(\delta)\, c - \frac{\Lambda_H}{h_1(H,T)} \right) & \frac{h_1(k^*,T)}{\sqrt{G(k^*)}} \\
&= \frac{\mathrm{sgn}(\delta)\, c\, h_1(k^*,T)}{\sqrt{G(k^*)}} - \frac{\delta\, g(k^*)\, r(w)}{\sigma\, \sqrt{G(k^*)}}\, \frac{h_1(k^*,T)}{h_1(H,T)} \\
&\quad + \mathcal{O}\big(w^{-\gamma}m^{-1/2} + m^{-\lambda-3/2}w^2 + m^{-1/2}\big) \\
&= \frac{\mathrm{sgn}(\delta)\, c\, h_1(k^*,T)}{\sqrt{G(k^*)}} - \frac{\delta\, g(k^*)\, r(w)}{\sigma\, \sqrt{G(k^*)}} + o(1),
\end{aligned}
\tag{2.6.16}
$$

where we already made use of the following four asymptotic relations:

1. Since $w \to \infty$, as $m \to \infty$ we have $w^{-\gamma}m^{-1/2} = o(1)$.

2. By our assumptions on ρ we have

$$m^{-\lambda-3/2}\, w^2 = \mathcal{O}\big(m^{-\lambda-3/2+2\beta+\rho(1-2\beta)}\big) = o(1).$$

3. Making use of (2.6.12) and our assumptions on ρ, we have

$$
\begin{aligned}
\frac{r(w)}{\sqrt{G(k^*)}} &\left(1 - \frac{h_1(k^*,T)}{h_1(H,T)}\right) \\
&= \mathcal{O}\big(w^2/m^{3/2}\big) \\
&= \mathcal{O}\big(m^{\rho(1-2\beta)+2\beta-3/2}\big) \\
&= o(1).
\end{aligned}
$$

4. Obviously it holds that $m^{-1/2} \to 0$.

Plugging the definition of w into (2.6.16) yields the assertion. $\square$

The following lemma allows us to neglect the time period before the change point in our asymptotic considerations.

Lemma 2.6.9. *Under the assumptions of Theorem 2.6.1 we have, as $m \to \infty$,*

$$A(\tilde{m}, k^*) \xrightarrow{P} -\infty,$$

where $A(a,b)$ is defined in (2.6.9).

Proof of Lemma 2.6.9. Let $1 < \tilde{\rho} < \rho$. By Theorem 2.4.1 we have

$$
\max_{k=\tilde{m},\ldots,k^*} \frac{|R_k|}{h_1(k,T)}
$$

$$
\leq \max_{k=\tilde{m},\ldots,k^*} \frac{|R_k|}{h_1(k,m^{\tilde{\rho}})} \max_{k=\tilde{m},\ldots,k^*} \frac{h_1(k,m^{\tilde{\rho}})}{h_1(k,T)}
$$

$$
= \mathcal{O}_P(1)\, o(1)
$$

$$
= o_P(1).
$$

Combining this with $\operatorname{sgn}(\delta)\, \Lambda_H/(h_1(H,T)) \to c > 0$ (see (2.6.13)) and

$$
h_1(k^*,T)/\sqrt{G(k^*)} = (G(T/\mu_0)/G(k^*))^{1/2-\beta} \to \infty,
$$

yields the assertion. $\qquad\square$

To shorten the upcoming proofs we state another technical lemma, showing that after the change point, the detectors are (asymptotically) driven by their proportion up to k^* and their deterministic perturbation.

Lemma 2.6.10. *Under the assumptions of Theorem 2.6.1 we have, as $m \to \infty$,*

$$
\max_{k=k^*+1,\ldots,H} \frac{|R_k - R_{k^*} - \Lambda_k|}{h_1(k,T)} = o_P\big(\sqrt{G(k^*)}/h_1(k^*,T)\big).
$$

Proof of Lemma 2.6.10. Recall that $M(k) = \sum_{i=1}^{k} \mu(i)$ denotes the mean of S_k (see Section 2.1). Via the approximation of Remark 1.3.3 we have for all $k = k^* + 1,\ldots,H$ that

$$
\left| \frac{R_k - R_{k^*} - \Lambda_k}{h_1(k,T)} \right|
$$

$$
= \left| \sum_{i=k^*+1}^{k} \frac{g(i)(\tilde{S}_i - \tilde{S}_{i-1} - \mu_0)}{h_1(k,T)} - \frac{\Lambda_k}{h_1(k,T)} \right|
$$

$$
= \left| \sum_{i=k^*+1}^{k} \frac{g(i)\big(\tilde{S}_i - M(i) - (\tilde{S}_{i-1} - M(i-1))\big)}{\sigma\, h_1(k,T)} \right|
$$

$$
\leq \frac{g(k)\,|\tilde{S}_k - M(k) - (\tilde{S}_{k^*} - M(k^*))|}{\sigma\, h_1(k,T)}
$$

$$
+ 2\frac{g(k) - g(k^*)}{\sigma\, h_1(k,T)} \max_{i=k^*,\ldots,k} |\tilde{S}_i - M(i)|.
$$

On the one hand, by the mean value theorem we have

$$
\max_{k=k^*+1,\ldots,H} |g(k) - g(k^*)| = \mathcal{O}\big(w/k^{*\lambda+1}\big)
$$

and on account of Corollary 2.3.3 and Lemma 2.6.8 it holds that

$$
\max_{i=k^*,\ldots,H} |\tilde{S}_i - M(i)| \overset{\text{a.s.}}{=} \mathcal{O}\big(\sqrt{k^* \log\log(k^*)}\big).
$$

Combining the two relations yields

$$\max_{k=k^*+1,\ldots,H} \frac{g(k) - g(k^*)}{h_1(k,T)} \max_{i=k^*+1,\ldots,k} |\tilde{S}_i - M(i)|$$
$$\overset{\text{a.s.}}{=} \mathcal{O}\big(\sqrt{\log\log(k^*)}\, k^{*-\lambda-1/2} w/h_1(k^*,T)\big).$$

On the other hand, we have via Theorem 2.3.2 that

$$\max_{k=k^*+1,\ldots,H} \frac{|\tilde{S}_k - M(k) - (\tilde{S}_{k^*} - M(k^*))|}{\sigma}$$

$$\overset{\text{a.s.}}{=} \max_{k=k^*+1,\ldots,H} \frac{|V(k) - M(k) - (V(k^*) - M(k^*))|}{\sigma} + \mathcal{O}\big(k^{*\kappa}\big) \qquad (2.6.17)$$

$$\leq 2 \max_{k=k^*,\ldots,H} \frac{|V(k) - \sigma\,W(k^*) - M(k)|}{\sigma} + \mathcal{O}\big(k^{*\kappa}\big),$$

where $V(k) = \max_{l=1,\ldots,k}\big(\sigma\,W(l)+M(l)\big)$ (see (2.3.2)). For each $k = k^*,\ldots,H$ we squeeze the upper expression into the following two bounds: On setting $l = k$ in the maximum of $V(k)$ we obtain as a lower bound

$$\big(V(k) - \sigma\,W(k^*) - M(k)\big)/\sigma \geq W(k) - W(k^*).$$

On establishing an upper bound we consider a sequence $\varsigma = \varsigma_m \in (0,1)$ with $\varsigma_m \to 0$ but $\varsigma_m \sqrt{k^*} \to \infty$. Taking into account the fact that $M(k)$ in increasing we obtain as an upper bound:

$$\big(V(k) - \sigma\,W(k^*) - M(k)\big)/\sigma$$

$$\leq \left(\frac{\max_{l=1,\ldots,[k^*(1-\varsigma)]}\big(\sigma\,W(l)+M(l)-\sigma\,W(k^*)-M(k)\big)}{\sigma}\right)_+$$

$$+ \left(\frac{\max_{l=[k^*(1-\varsigma)]+1,\ldots,k}\big(\sigma\,W(l)+M(l)-\sigma\,W(k^*)-M(k)\big)}{\sigma}\right)_+$$

$$\leq Z(k^*)_+ + \max_{l=[k^*(1-\varsigma)]+1,\ldots,k} |W(l) - W(k^*)|,$$

where

$$Z(k^*) := \max_{l=1,\ldots,[k^*(1-\varsigma)]}\big(W(l) - W(k^*)\big) + \frac{M([k^*(1-\varsigma)]) - M(k^*)}{\sigma}$$

$$= \max_{l=1,\ldots,[k^*(1-\varsigma)]}\big(W(l) - W(k^*)\big) - \sum_{i=[k^*(1-\varsigma)]+1}^{k^*} \mu(i)/\sigma$$

$$\leq \max_{l=1,\ldots,[k^*(1-\varsigma)]}\big(W(l) - W(k^*)\big) - k^*\,\varsigma\,\min\{\mu_0,\mu_1\}/\sigma$$

and, by the fact that $W(t + k^*) - W(k^*)$ is again a Wiener process,

$$\max_{l=[k^*(1-\varsigma)]+1,\ldots,k} |W(l) - W(k^*)|$$

$$\leq \max_{l=[k^*(1-\varsigma)]+1,\ldots,k^*} |W(l) - W(k^*)| + \max_{l=k^*+1,\ldots,k} |W(l) - W(k^*)|$$

$$= \mathcal{O}_P\big(\sqrt{\varsigma k^*}\big) + \mathcal{O}_P\big(\sqrt{H - k^*}\big).$$

Combining the upper and lower bound yields

$$
\begin{aligned}
\max_{k=k^*,\dots,H} \frac{|V(k) - \sigma\, W(k^*) - M(k)|}{\sigma} \\
\leq\ Z(k^*)_+ + \mathcal{O}_P\big(\sqrt{\varsigma k^*}\big) + \mathcal{O}_P\big(\sqrt{H-k^*}\big) \\
=\ \mathcal{O}_P(1) + \mathcal{O}_P\big(\sqrt{\varsigma k^*}\big) + \mathcal{O}_P(\sqrt{w}),
\end{aligned}
$$

where we already plugged in the fact that

$$
\begin{aligned}
P(|Z(k^*)_+| > 0) &= P(Z(k^*) > 0) \\
&\leq P\big(\mathcal{O}_P(\sqrt{k^*}) - k^*\,\varsigma\,\min\{\mu_0,\mu_1\}/\sigma > 0\big) \\
&= P\big(\mathcal{O}_P(1) - \sqrt{k^*}\,\varsigma\,\min\{\mu_0,\mu_1\}/\sigma > 0\big) \\
&\to\ 0,
\end{aligned}
$$

since $\sqrt{k^*}\,\varsigma \to \infty$. In conclusion we have

$$
\begin{aligned}
\max_{k=k^*+1,\dots,H} \frac{|R_k - R_{k^*} - \Lambda_k|}{h_1(k,T)} \\
=\ \mathcal{O}_P\big(\sqrt{\log\log(k^*)}\,k^{*-\lambda-1/2}w/h_1(k^*,T)\big) + \mathcal{O}_P\big(k^{*\kappa}/h_1(k^*,T)\big) \\
+\ \mathcal{O}_P\big(\sqrt{k^*\varsigma_m}/h_1(k^*,T)\big) + \mathcal{O}_P\big(\sqrt{w}/h_1(k^*,T)\big) \\
=\ o_P\big(\sqrt{G(k^*)}/h_1(k^*,T)\big),
\end{aligned}
\tag{2.6.18}
$$

where the last equation is a consequence of the following implications of Lemma 2.6.8 and our assumption on ρ:

1. $k^{*-\lambda-1/2}w/\sqrt{G(k^*)} \simeq m^{-\lambda-1/2+(\rho-1)(1/2-\beta)} \to 0$,

2. $k^{*\kappa}/\sqrt{G(k^*)} \simeq k^{*\kappa-1/2} \to 0$,

3. $\sqrt{k^*\varsigma_m/G(k^*)} \simeq \sqrt{\varsigma_m} \to 0$,

4. $\sqrt{w/G(k^*)} \simeq \sqrt{w/k^*} \to 0$.

$\square$

The next lemma allows us to restrict the range of the maximum to an area arbitrary close to H.

Lemma 2.6.11. Let $\epsilon > 0$ and $\check{H} = \check{H}(\epsilon) = k^* + \big[(1-\epsilon)w\big]$. Under the assumptions of Theorem 2.6.1 it holds that, as $m \to \infty$,

$$
P\big(A(k^*+1,H) = A(\check{H},H)\big) \to 1.
$$

Proof of Lemma 2.6.11. The proof follows essentially by the fact that $\Lambda_H - \Lambda_{\check{H}} \to \infty$ sufficiently fast. For the sake of simplicity we demonstrate the proof for $\delta > 0$. The case of $\delta < 0$ follows by the same arguments. The first step of the proof is to show that is sufficient to consider the (deterministic) perturbation of the detectors in the respective range:

$$
\begin{aligned}
P\big(A(k^* + 1, H) &= A(\check{H}, H)\big) \\
&= P\big(A(k^* + 1, \check{H} - 1) \le A(\check{H}, H)\big) \\
&\ge P\big(A(k^* + 1, \check{H} - 1) \le A(H, H)\big) \\
&= P\left(\max_{k=k^*+1,\ldots,\check{H}-1} \frac{R_{k^*} + R_k - R_{k^*}}{h_1(k, T)} \le \frac{R_H}{h_1(H, T)}\right) \qquad (2.6.19) \\
&\ge P\left(\max_{k=k^*+1,\ldots,\check{H}-1} \frac{R_{k^*}}{h_1(k, T)} + \max_{k=k^*+1,\ldots,\check{H}-1} \frac{R_k - R_{k^*}}{h_1(k, T)} \le \frac{R_H}{h_1(H, T)}\right) \\
&= P\left(\max_{k=k^*+1,\ldots,\check{H}-1} \frac{R_k - R_{k^*}}{h_1(k, T)} \le \frac{R_H - R_{k^*}}{h_1(H, T)} + o_P(1)\right),
\end{aligned}
$$

where in the last line we already plugged in that by Lemma 2.6.8 we have

$$
\begin{aligned}
\bigg|\max_{k=k^*+1,\ldots,\check{H}-1} &\frac{R_{k^*}}{h_1(k, T)} - \frac{R_{k^*}}{h_1(H, T)}\bigg| \\
&\le \frac{|R_{k^*}|}{h_1(k^*, T)}\left(\max_{k=k^*+1,\ldots,\check{H}-1} \frac{h_1(k^*, T)}{h_1(k, T)} - \frac{h_1(k^*, T)}{h_1(H, T)}\right) \qquad (2.6.20) \\
&= \mathcal{O}_P(1)\left(1 - \frac{h_1(k^*, T)}{h_1(H, T)}\right) \\
&= o_P(1).
\end{aligned}
$$

By Lemma 2.6.10 we know that the stochastic terms of (2.6.19) are asymptotically negligible, hence

$$
\begin{aligned}
P\big(A(k^* + 1, H) &= A(\check{H}, H)\big) \\
&\ge P\left(\max_{k=k^*+1,\ldots,\check{H}-1} \frac{\Lambda_k}{h_1(k, T)} \le \frac{\Lambda_H}{h_1(H, T)} + o_P(1)\right) \\
&= P\left(\max_{k=k^*+1,\ldots,\check{H}-1} \frac{\Lambda_k}{G(k)^\beta} \le \frac{\Lambda_H}{G(H)^\beta} + o_P\big(T^{1/2-\beta}\big)\right).
\end{aligned}
$$

By (2.6.13) and the fact that Λ_k is positive and increasing for $k > k^*$ we have uniformly for $k = k^* + 1, \ldots, \check{H} - 1$

$$
\max_k \frac{\Lambda_k}{G(k)^\beta} \le \max_k \frac{\Lambda_k}{G(H)^\beta}\left(\frac{G(H)}{G(k^*)}\right)^\beta = \frac{\Lambda_{\check{H}-1}}{G(H)^\beta}(1 + o(1)),
$$

which gives us

$$
\begin{aligned}
P\bigg(\max_{k=k^*+1,\ldots,\check{H}-1} &\frac{\Lambda_k}{G(k)^\beta} \le \frac{\Lambda_H}{G(H)^\beta} + o_P\big(T^{1/2-\beta}\big)\bigg) \\
&\ge P\left(0 \le \frac{\Lambda_H}{G(H)^\beta} - \frac{\Lambda_{\check{H}-1}}{G(H)^\beta}(1 + o(1)) + o_P\big(T^{1/2-\beta}\big)\right) \qquad (2.6.21) \\
&= P\big(0 \le \Lambda_H - \Lambda_{\check{H}-1} + o(\Lambda_{\check{H}-1}) + o_P(h_1(H, T))\big).
\end{aligned}
$$

Now, on the one hand, it holds that

$$\frac{\Lambda_H - \Lambda_{\check{H}-1}}{w}$$

$$= \frac{\delta}{\sigma\,w} \sum_{i=k^*+[(1-\epsilon)w]}^{k^*+[w]} g(i)\big(1 - (1 + i - k^*)^{-\gamma}\big)$$

$$\geq \frac{\delta}{\sigma} g(k^* + [(1-\epsilon)w]) \left(1 - (1 + [(1-\epsilon)w])^{-\gamma}\right) \frac{\epsilon\,[w]}{w}$$

$$\to \frac{\delta\,\epsilon}{\sigma}.$$

On the other hand, we have $\Lambda_{\check{H}-1} = \mathcal{O}(w)$ and $w \simeq h_1(H,T)$. Plugging this in (2.6.21) yields

$$P\Big(0 \leq \Lambda_H - \Lambda_{\check{H}-1} + o\big(\Lambda_{\check{H}-1}\big) + o_P(h_1(H,T))\Big)$$

$$\geq P(0 \leq \delta\,\epsilon + o_P(1))$$

$$\to 1,$$

which completes the proof. $\qquad\square$

The following lemma gives us the required convergence towards the standard normal distribution.

Lemma 2.6.12. *Under the assumptions of Theorem 2.6.1 we have under* $H_{1,+}^{closed}$ *respectively* $H_{1,-}^{closed}$ *that*

$$\lim_{m\to\infty} P\big(\tau_{m,+}^{closed} \leq H\big) = \phi(x),$$

$$\lim_{m\to\infty} P\big(\tau_{m,-}^{closed} \leq H\big) = \phi(x).$$

Proof of Lemma 2.6.12. We consider the one-sided, positive stopping time $\tau_{m,+}^{closed}$ (i.e. the case $\delta > 0$) in detail: On combining Lemma 2.6.8, Lemma 2.6.9 and Lemma 2.6.11 we have

$$P\big(\tau_{m,+}^{closed} \leq H\big)$$

$$= P(A(\tilde{m}, H) > -x + o(1))$$

$$= P(A(k^* + 1, H) > -x + o(1)) + o(1) \qquad (2.6.22)$$

$$= P(A(\check{H}, H) > -x + o(1)) + o(1).$$

We squeeze $A(\check{H}, H)$ between the following bounds:

$$\frac{h_1(k^*,T)}{\sqrt{G(k^*)}}\left(\frac{R_H}{h_1(H,T)} - \frac{\Lambda_H}{h_1(H,T)}\right)$$

$$\leq A(\check{H}, H)$$

$$= \frac{h_1(k^*,T)}{\sqrt{G(k^*)}} \max_{k=\check{H},\dots,H}\left(\frac{R_k}{h_1(k,T)} - \frac{\Lambda_H}{h_1(H,T)}\right) \qquad (2.6.23)$$

$$\leq \frac{h_1(k^*,T)}{\sqrt{G(k^*)}} \max_{k=\check{H},\dots,H}\left(\frac{R_k}{h_1(k,T)} - \frac{\Lambda_k}{h_1(k,T)}\right) + o(1),$$

where in the last line we used, for one thing, Λ_k being increasing for $k > k^*$ and, for another thing, a combination of Lemma 2.6.8, the early change assumption (2.6.2) and our assumptions on ρ, i.e.

$$
\begin{aligned}
\frac{h_1(k^*,T)}{\sqrt{G(k^*)}} &\max_{k=\check{H},\dots,H} \left| \frac{\Lambda_H}{h_1(k,T)} - \frac{\Lambda_H}{h_1(H,T)} \right| \\
&\leq \frac{h_1(k^*,T)}{\sqrt{G(k^*)}} \frac{|\mu_1 - \mu_0|\, w}{h_1(H,T)} \max_{k=\check{H},\dots,H} \left| \frac{h_1(H,T)}{h_1(k,T)} - 1 \right| \\
&= \mathcal{O}(1) \frac{h_1(k^*,T)}{\sqrt{G(k^*)}} \frac{w}{m} \\
&= o(1).
\end{aligned}
$$

Applying Lemma 2.6.10 to the lower and upper bound of (2.6.23) yields

$$
\begin{aligned}
\frac{h_1(k^*,T)}{\sqrt{G(k^*)}} &\frac{R_{k^*}}{h_1(H,T)} + o_P(1) \\
&\leq A(\check{H},H) \\
&\leq \frac{h_1(k^*,T)}{\sqrt{G(k^*)}} \max_{k=\check{H},\dots,H} \frac{R_{k^*}}{h_1(k,T)} + o_P(1),
\end{aligned}
$$

where in a last step we confirm that we can replace the argument in the threshold function by k^*: Making use of Lemma 2.6.8 and the assumption on ρ, once more, we obtain

$$
\begin{aligned}
\frac{h_1(k^*,T)}{\sqrt{G(k^*)}} &\left(\max_{k=\check{H},\dots,H} \frac{R_{k^*}}{h_1(k,T)} - \frac{R_{k^*}}{h_1(k^*,T)} \right) \\
&\leq \frac{h_1(k^*,T)}{\sqrt{G(k^*)}} \frac{|R_{k^*}|}{h_1(k^*,T)} \max_{k=\check{H},\dots,H} \left| \frac{h_1(k^*,T)}{h_1(k,T)} - 1 \right| \\
&= \frac{h_1(k^*,T)}{\sqrt{G(k^*)}} \mathcal{O}_P(1)\, \mathcal{O}(w/m) \\
&= o_P(1).
\end{aligned}
$$

Hence, the upper and lower bound are asymptotically equal and we obtain

$$
A(\check{H},H) = \frac{R_{k^*}}{\sqrt{G(H^*)}} + o_P(1).
$$

Along the lines of the proof of Theorem 2.4.1 we have

$$
\frac{R_{k^*}}{\sqrt{G(k^*)}} = \frac{\int_0^{k^*} g(x)\,dW(x)}{\sqrt{G(k^*)}} + o_P(1)
$$

and $\int_0^{k^*} g(x)\,dW(x)/\sqrt{G(k^*)} \stackrel{D}{=} W(1)$. Hence, by (2.6.22) we have

$$
P\big(\tau_{m,+}^{closed} \leq H\big) = P(W(1) \geq -x + o(1)) + o(1) \to \phi(x),
$$

which completes the proof of the assertion for the one-sided, positive stopping time τ_+^{closed}. We briefly outline the analogue steps on proving the assertion for the one-sided, negative stopping time: Under $H_{1,-}^{closed}$ we have

$$
\begin{aligned}
P\big(\tau_{m,-}^{closed} &\leq H\big) \\
&= P\left(\max_{k=\tilde{m},\dots,H} \frac{-R_k}{h_1(k,T)} > c \right) \\
&= P\left(\frac{-\int_0^{k^*} g(x)\, dW(x)}{\sqrt{G(k^*)}} + o_P(1) > \left(c + \frac{\Lambda_H}{h_1(H,T)} \right) \frac{h_1(k^*,T)}{\sqrt{G(k^*)}} \right) \\
&= P\left(\frac{-\int_0^{k^*} g(x)\, dW(x)}{\sqrt{G(k^*)}} + o_P(1) > -x + o(1) \right) \\
&\to P\big(W(1) < x \big) \\
&= \phi(x).
\end{aligned}
$$
(2.6.24)

$\square$

Finally, we present the proof of the first main theorem of this section.

Proof of Theorem 2.6.1. Lemma 2.6.12 gives us the required convergence towards the standard normal distribution. So, it just remains to solve $\tau_{m,+}^{closed} \leq H(x) = k^* + [w(k^*)]$ for x and to replace the unknown k^* by $\tilde{m}$ (which is possible by the early-change assumption). By Lemma 2.6.8 we have $w \to \infty$, so it holds for sufficiently large m that

$$
\begin{aligned}
P\big(\tau_{m,+}^{closed} &\leq H\big) \\
&= P\big(\tau_{m,+}^{closed} - k^* \leq [w]\big) \\
&= P\big(\max\{\tau_m^{closed} - k^*, 1\} \leq [w]\big).
\end{aligned}
$$

Thus, we can apply the increasing function r on both sides of the inequality and obtain

$$
P\big(\tau_{m,+}^{closed} \leq H\big) = P\Big(r\big(\max\{\tau_{m,+}^{closed} - k^*, 1\} \big) \leq r([w(k^*)]) \Big).
$$
(2.6.25)

On replacing the unknown k^* by $\tilde{m}$ we aim to show that

$$
r([w(k^*)]) \stackrel{\text{a.s.}}{=} r(w(\tilde{m})) + o(\sqrt{m}).
$$
(2.6.26)

An application of the mean value theorem shows that $r([w(k^*)]) = r(w(k^*)) + \mathcal{O}(1)$. On comparing $r(w(k^*))$ and $r(w(\tilde{m}))$ we have, for one thing, that for some

$\xi \in [\tilde{m}, k^*]$

$$\left| \frac{\sqrt{G(k^*)}}{g(k^*)} - \frac{\sqrt{G(\tilde{m})}}{g(\tilde{m})} \right|$$

$$\leq \left| \frac{\sqrt{G(k^*)} - \sqrt{G(\tilde{m})}}{g(k^*)} \right| + \sqrt{G(\tilde{m})} \left| \frac{1}{g(k^*)} - \frac{1}{g(\tilde{m})} \right|$$

$$\overset{\text{a.s.}}{=} \frac{1}{2} \frac{g^2(\xi)}{\sqrt{G(\xi)}} \, |k^* - \tilde{m}| + \mathcal{O}(\sqrt{m}) \, o(1) \qquad (2.6.27)$$

$$\overset{\text{a.s.}}{=} \mathcal{O}((k^* - \tilde{m})/\sqrt{m}) + o(\sqrt{m})$$

$$\overset{\text{a.s.}}{=} o(\sqrt{m})$$

and, for another thing, by the early-change assumption and a repeated application of the mean value theorem that for $0 < \beta < 1/2$

$$\left| \frac{h_1(k^*, T)}{g(k^*)} - \frac{h_1(\tilde{m}, T)}{g(\tilde{m})} \right|$$

$$= \frac{|h_1(k^*, T) - h_1(\tilde{m}, T)|}{g(k^*)} + h_1(\tilde{m}, T) \left| \frac{1}{g(k^*)} - \frac{1}{g(\tilde{m})} \right|$$

$$\overset{\text{a.s.}}{=} \mathcal{O}(T^{1/2-\beta}) \, |(G(k^*))^\beta - (G(\tilde{m}))^\beta|$$

$$\quad + \mathcal{O}(T^{1/2-\beta} m^\beta \, |k^* - \tilde{m}| \, m^{-\lambda-1})$$

$$\overset{\text{a.s.}}{=} \mathcal{O}(T^{1/2-\beta} |k^* - \tilde{m}| \, m^{\beta-1}) + \mathcal{O}(T^{1/2-\beta} m^{\beta+\eta-\lambda-1})$$

$$\overset{\text{a.s.}}{=} \mathcal{O}(T^{1/2-\beta} m^{\beta+\eta-1})$$

$$= \mathcal{O}(m^{(\rho-1)(1/2-\beta)+\eta-1/2})$$

$$= o(\sqrt{m}).$$

For $\beta = 0$ the first term in the second line disappears, yet we obtain the same rate of convergence. Hence, (2.6.26) holds true and therefore we have

$$P\left(\frac{|\delta| \, g(\tilde{m}) \, r\big(\max\{\tau_{m,+}^{closed} - k^*, 1\}\big)/\sigma - c \, h_1(\tilde{m}, T)}{\sqrt{G(\tilde{m})}} \leq x \right) \to \phi(x).$$

The proof is complete if we replace δ by its estimate $\hat{\delta} = \delta + o_P(m^{(\rho-1)(1/2-\beta)})$. Denoting $X_m := g(\tilde{m}) \, r\big(\max\{\tau_{m,+}^{closed} - k^*, 1\}\big)/(\sigma \sqrt{G(\tilde{m})})$ we find that

$$|\hat{\delta}| X_m - \frac{c \, h_1(\tilde{m}, T)}{\sqrt{G(\tilde{m})}}$$

$$= (|\hat{\delta}|/|\delta|) \, |\delta| X_m - \frac{c \, h_1(\tilde{m}, T)}{\sqrt{G(\tilde{m})}}$$

$$= |\delta| X_m - \frac{c \, h_1(\tilde{m}, T)}{\sqrt{G(\tilde{m})}} + o_P\left(m^{(1-\rho)(1/2-\beta)} X_m \right)$$

$$= |\delta| X_m - \frac{c \, h_1(\tilde{m}, T)}{\sqrt{G(\tilde{m})}} + o_P(m^{(1-\rho)(1/2-\beta)} h_1(\tilde{m}, T)/\sqrt{G(\tilde{m})}),$$

$$= |\delta| X_m - \frac{c \, h_1(\tilde{m}, T)}{\sqrt{G(\tilde{m})}} + o_P(1),$$

which yields the first assertion. The same arguments hold true for $\tau_{m,-}^{closed}$, hence the proof is complete. $\qquad\square$

Proof of Theorem 2.6.2

Lemma 2.6.13. *Under the assumptions of Theorem 2.6.2 we have, as* $m \to \infty$,

$$w \overset{a.s.}{\simeq} h_2(k^*, \tilde{m}) \overset{a.s.}{\longrightarrow} \infty, \tag{2.6.28}$$

$$H \overset{a.s.}{\sim} k^* \sim m/\mu_0, \tag{2.6.29}$$

$$G(H)/G(k^*) \overset{a.s.}{=} 1 + \mathcal{O}(w/k^*) \tag{2.6.30}$$

$$h_2(H, \tilde{m})/h_2(k^*, \tilde{m}) \overset{a.s.}{=} 1 + \mathcal{O}(w/k^*) \tag{2.6.31}$$

$$\left(\operatorname{sgn}(\delta) - \frac{\Lambda_H}{h_2(H, \tilde{m})} \right) \frac{h_2(H, \tilde{m})}{\sqrt{G(k^*)}} \overset{a.s.}{\longrightarrow} -\operatorname{sgn}(\delta)\, x. \tag{2.6.32}$$

Proof of Lemma 2.6.13. Recall that

$$w = r^{-1} \left(\frac{\sqrt{G(k^*)}\, \sigma}{g(k^*)\, |\delta|} \left(\frac{h_2(k^*, \tilde{m})}{\sqrt{G(k^*)}} + x \right) \right).$$

By the non-contamination assumption (2.1.3) it holds that

$$\frac{h_2(k^*, \tilde{m})}{\sqrt{G(k^*)}} \geq f^{-1}(\ln(G(m/\mu_0)/G(\tilde{m})) + f(c)) \overset{a.s.}{\longrightarrow} c,$$

so, by $x > -c$ the sequence $h_2(k^*, \tilde{m})/\sqrt{G(k^*)} + x$ is a.s. bounded away from zero. Combining this with $\sqrt{G(k^*)}/g(k^*) \to \infty$ and Lemma 2.6.6 yields the first assertion. The second one can be seen as follows: By Remark 1.7.1, Lemma 2.2.3 and (2.6.28), we have

$$\frac{H - k^*}{k^*} \overset{a.s.}{\simeq} \frac{h_2(k^*, \tilde{m})}{k^*} \sim \sqrt{\frac{G(k^*)}{k^*}} \sqrt{\frac{\ln(G(k^*)/G(\tilde{m})) + f(c)}{k^*}} \overset{a.s.}{\longrightarrow} 0.$$

Assertion (2.6.30) is shown in Lemma 2.6.8 and just restated here for the sake of completeness. Concerning (2.6.31) we first note that by the mean value theorem we have for some $\xi \in [G(k^*)/G(\tilde{m}) + f(c), G(H)/G(\tilde{m}) + f(c)]$, that

$$\frac{f^{-1}(\ln(G(H)/G(\tilde{m})) + f(c))}{f^{-1}(\ln(G(k^*)/G(\tilde{m})) + f(c))} - 1$$

$$= \left(f^{-1}(\xi) \right)' \frac{\ln(G(H)/G(\tilde{m})) - \ln(G(k^*)/G(\tilde{m}))}{f^{-1}(\ln(G(k^*)/G(\tilde{m})) + f(c))},$$

where, on the one hand, by (2.4.16) $(f^{-1}(\xi))' = \mathcal{O}(1)$ and, on the other hand, by a further application of the mean value theorem for some $\tilde{\xi} \in [1, H/k^*]$

$$\frac{\ln(G(H)/G(\tilde{m})) - \ln(G(k^*)/G(\tilde{m}))}{f^{-1}(\ln(G(k^*)/G(\tilde{m})) + f(c))}$$

$$\leq \frac{\ln(G(H)/G(k^*))}{f^{-1}(\ln(G(m/\mu_0)/G(\tilde{m})) + f(c))}$$

$$= \frac{G(H)/G(k^*) - 1}{c\,\tilde{\xi}\, f^{-1}(\ln(G(m/\mu_0)/G(\tilde{m})) + f(c))}$$

$$\overset{a.s.}{=} \mathcal{O}(w/k^*).$$

Combining this with (2.6.30) yields

$$\frac{h_2(H, \tilde{m})}{h_2(k^*, \tilde{m})} = \sqrt{\frac{G(H)}{G(k^*)}} \frac{f^{-1}(\ln(G(H)/G(\tilde{m})) + f(c))}{f^{-1}(\ln(G(k^*)/G(\tilde{m})) + f(c))}$$
$$\overset{\text{a.s.}}{=} (1 + \mathcal{O}(w/k^*)) (1 + \mathcal{O}(w/k^*))$$
$$= 1 + \mathcal{O}(w/k^*).$$

So, (2.6.32) remains to be shown. By (2.6.15) we have

$$\Lambda_H \overset{\text{a.s.}}{=} \frac{\delta\, g(k^*)\, r(w)}{\sigma} + \mathcal{O}\big(w^{-\gamma} + m^{-\lambda-1}w^2 + 1\big),$$

which yields

$$\left(\text{sgn}(\delta) - \frac{\Lambda_H}{h_2(H, \tilde{m})} \right) \frac{h_2(H, \tilde{m})}{\sqrt{G(k^*)}}$$
$$\overset{\text{a.s.}}{=} \frac{\text{sgn}(\delta)\, h_2(H, \tilde{m})}{\sqrt{G(k^*)}} - \frac{\delta\, g(k^*) r(w)}{\sigma\, \sqrt{G(k^*)}}$$
$$+ \mathcal{O}\big(w^{-\gamma}\, m^{-1/2} + m^{-\lambda-3/2}\, w^2 + m^{-1/2}\big)$$
$$= \text{sgn}(\delta)\left(\frac{h_2(H, \tilde{m})}{\sqrt{G(k^*)}} - \frac{h_2(k^*, \tilde{m})}{\sqrt{G(k^*)}} - x \right) + o(1)$$
$$= \text{sgn}(\delta)\left(\frac{h_2(H, \tilde{m})}{\sqrt{G(k^*)}}\left(1 - \frac{h_2(k^*, \tilde{m})}{h_2(H, \tilde{m})} \right) - x \right) + o(1)$$
$$\overset{\text{a.s.}}{=} -\text{sgn}(\delta)\, x + o(1),$$

and therefore completes the proof. $\square$

Lemma 2.6.14. *Under the assumptions of Theorem 2.6.2 we have*

$$\max_{k=\tilde{m},\ldots,H} \frac{|R_k - R_{k^*} - \Lambda_k|}{h_2(k, \tilde{m})} = o_P(1).$$

Proof of Lemma 2.6.14. We consider the maximum for $k \leq k^*$ and $k > k^*$ separately. Note that before the change point $\Lambda_k = 0$. By the invariance principle of Theorem 2.3.1 we have

$$\max_{k=\tilde{m},\ldots,k^*} \frac{|R_{k^*} - R_k|}{h_2(k, \tilde{m})}$$
$$\overset{\text{a.s.}}{=} \max_{k=\tilde{m},\ldots,k^*} \frac{|\sum_{i=k+1}^{k^*} g(i)(W(i) - W(i-1))|}{h_2(k, \tilde{m})} + \mathcal{O}\big(k^{*\kappa}/h_2(\tilde{m}, \tilde{m})\big),$$

where the latter sum can be approximated as shown in Remark 1.3.3:

$$\left| \sum_{i=k+1}^{k^*} g(i)(W(i) - W(i-1)) \right|$$
$$\leq g(k^*)\, |W(k^*) - W(k)| + 2\, |g(k^*) - g(k+1)| \max_{k \leq t \leq k^*} |W(t)|.$$

Making use of the fact that $W(t+k^*) - W(k^*)$ is again a Wiener process, the mean value theorem and the law of the iterated logarithm we obtain

$$\max_{k=\tilde{m},\ldots,k^*} \frac{|R_{k^*} - R_k|}{h_2(k,\tilde{m})}$$
$$= \mathcal{O}_P\big(k^{*\kappa}/h_2(\tilde{m},\tilde{m})\big)$$
$$\quad + \mathcal{O}_P\big(\sqrt{(k^* - \tilde{m})}/h_2(\tilde{m},\tilde{m})\big) \tag{2.6.33}$$
$$\quad + \mathcal{O}_P\big((k^* - \tilde{m})\sqrt{k^* \log\log(k^*)}/(m^{1+\lambda} h_2(\tilde{m},\tilde{m}))\big)$$
$$= \mathcal{O}_P\big(m^{\kappa-1/2}\big) + \mathcal{O}_P\big(m^{(\eta-1)/2}\sqrt{\log(m)}\big)$$
$$= o_P(1).$$

On the other hand, by replacing $h_1(k,T)$ by $h_2(k,\tilde{m})$ in Lemma 2.6.10 (cf. (2.6.18)), we obtain

$$\max_{k=k^*+1,\ldots,H} \frac{|R_k - R_{k^*} - \Lambda_k|}{h_2(k,\tilde{m})}$$
$$= \mathcal{O}_P\big(\sqrt{\log\log(k^*)}\, k^{*-\lambda-1/2}w/h_2(k^*,\tilde{m})\big) + \mathcal{O}_P\big(k^{*\kappa}/h_2(k^*,\tilde{m})\big),$$
$$\quad + \mathcal{O}_P\big(\sqrt{k^* \varsigma_m}/h_2(k^*,\tilde{m})\big) + \mathcal{O}_P\big(h_2(k^*,\tilde{m})^{-1/2}\big).$$
$$= o_P(1),$$

where the last equality follows from $w \overset{\text{a.s.}}{\simeq} h_2(\tilde{m},\tilde{m}) \overset{\text{a.s.}}{\simeq} \sqrt{m}.$ $\qquad \square$

Lemma 2.6.15. *Under the assumptions of Theorem 2.6.2 we have under $H_{1,+}^{open}$ respectively $H_{1,+}^{open}$ that*

$$\lim_{m\to\infty} P\big(\tau_{m,+}^{open} \leq H\big) = \phi(x),$$
$$\lim_{m\to\infty} P\big(\tau_{m,-}^{open} \leq H\big) = \phi(x).$$

Proof of Lemma 2.6.15. We consider the one-sided, positive stopping time $\tau_{m,+}^{open}$ (i.e. the case $\delta > 0$) in detail: By Lemma 2.6.14 we have

$$\max_{k=\tilde{m},\ldots,H} \frac{R_k}{h_2(k,\tilde{m})} - \frac{\Lambda_H}{h_2(H,\tilde{m})}$$
$$= \max_{k=\tilde{m},\ldots,H} \frac{R_{k^*} + \Lambda_k}{h_2(k,\tilde{m})} - \frac{\Lambda_H}{h_2(H,\tilde{m})} + o_P(1),$$

where we replace the right hand side by $R_{k^*}/h_2(H,\tilde{m})$ via the following three arguments:

1. By (2.6.31) we have

$$\max_{k=\tilde{m},\ldots,H} \left| \frac{R_{k^*}}{h_2(k,\tilde{m})} - \frac{R_{k^*}}{h_2(H,\tilde{m})} \right| = \mathcal{O}_P(w/k^*) = o_P(1).$$

2. Since $\Lambda_k = 0$ for $k \leq k^*$ but $\Lambda_k > 0$ for $k > k^*$ it holds that

$$\max_{k=\tilde{m},\ldots,H} \frac{\Lambda_k}{h_2(k,\tilde{m})} - \frac{\Lambda_k}{h_2(H,\tilde{m})}$$

$$= \max_{k=k^*+1,\ldots,H} \frac{\Lambda_k}{h_2(k,\tilde{m})} - \frac{\Lambda_k}{h_2(H,\tilde{m})}$$

and by (2.6.32) and (2.6.31) we obtain

$$\max_{k=k^*+1,\ldots,H} \left| \frac{\Lambda_k}{h_2(k,\tilde{m})} - \frac{\Lambda_k}{h_2(H,\tilde{m})} \right|$$

$$= \max_{k=k^*+1,\ldots,H} \frac{|\Lambda_k|}{h_2(H,\tilde{m})} \left(\frac{h_2(H,\tilde{m})}{h_2(k,\tilde{m})} - 1 \right)$$

$$\overset{\text{a.s.}}{=} \mathcal{O}\left(w/k^* \right)$$

$$\overset{\text{a.s.}}{=} o(1).$$

3. Since Λ_k is non-decreasing in k, it holds that $\max_{k=\tilde{m},\ldots,H}(\Lambda_k - \Lambda_H) = 0$.

The three assertions can be combined as follows:

$$\left| \max_{k=\tilde{m},\ldots,H} \frac{R_{k^*} + \Lambda_k}{h_2(k,\tilde{m})} - \frac{\Lambda_H}{h_2(H,\tilde{m})} - \frac{R_{k^*}}{h_2(H,\tilde{m})} \right|$$

$$= \left| \max_{k=\tilde{m},\ldots,H} \frac{R_{k^*} + \Lambda_k}{h_2(k,\tilde{m})} - \frac{R_{k^*} + \Lambda_H}{h_2(H,\tilde{m})} \right|$$

$$\leq \max_{k=\tilde{m},\ldots,H} \left| \frac{R_{k^*} + \Lambda_k}{h_2(k,\tilde{m})} - \frac{R_{k^*} + \Lambda_H}{h_2(H,\tilde{m})} \right| \tag{2.6.34}$$

$$\leq \max_{k=\tilde{m},\ldots,H} \left| \frac{R_{k^*}}{h_2(k,\tilde{m})} - \frac{R_{k^*}}{h_2(H,\tilde{m})} \right| + \left| \frac{\Lambda_k}{h_2(k,\tilde{m})} - \frac{\Lambda_H}{h_2(H,\tilde{m})} \right|$$

$$= o_P(1).$$

Thus, it holds true that

$$\max_{k=\tilde{m},\ldots,H} \frac{R_k}{h_2(k,\tilde{m})} - \frac{\Lambda_H}{h_2(H,\tilde{m})} = \frac{R_{k^*}}{h_2(H,\tilde{m})} + o_P(1)$$

and by the early-change assumption it holds that

$$\frac{h_2(H,\tilde{m})}{\sqrt{G(k^*)}} \overset{\text{a.s.}}{\simeq} f^{-1}(\ln(G(H)/G(\tilde{m})) + f(c)) \overset{\text{a.s.}}{=} \mathcal{O}(1).$$

Combining the assertions above gives us

$$\frac{h_2(H,\tilde{m})}{\sqrt{G(k^*)}} \max_{k=\tilde{m},\ldots,H} \frac{R_k}{h_2(k,\tilde{m})} - \frac{\Lambda_H}{h_2(H,\tilde{m})} = \frac{R_{k^*}}{\sqrt{G(k^*)}} + o_P(1).$$

Along the lines of the proof of Theorem 2.4.2 we make use of the invariance principle to replace our observations by a Wiener process. The Gaussian analogue of the detectors can be approximated by a Wiener integral which is again distributed as a Wiener process. Combining this with (2.6.32) gives us

$$P\big(\tau_{m,+}^{open} \le H\big) \ \to \ P(W(1) > -x) \ = \ \phi(x).$$

The same arguments yield in case of $\delta < 0$ that

$$\frac{h_2(H,\tilde{m})}{\sqrt{G(k^*)}} \left(\max_{k=\tilde{m},\dots,H} \frac{-R_k}{h_2(k,\tilde{m})} + \frac{\Lambda_k}{h_2(k,\tilde{m})} \right) \ = \ \frac{-R_{k^*}}{\sqrt{G(k^*)}} + o_P(1),$$

hence

$$
\begin{aligned}
P\big(\tau_{m,-}^{open} &\le H\big) \\
&= P\Big(\max_{k=\tilde{m},\dots,H} -R_k/h_2(k,\tilde{m}) > 1 \Big) \\
&= P\left(-W(1) > \left(1 + \frac{\Lambda_H}{h_2(H,\tilde{m})}\right) \frac{h_2(H,\tilde{m})}{\sqrt{G(k^*)}} \right) + o(1) \\
&= P\left(-W(1) > -\left(\mathrm{sgn}(\delta) - \frac{\Lambda_H}{h_2(H,\tilde{m})}\right) \frac{h_2(H,\tilde{m})}{\sqrt{G(k^*)}} \right) + o(1) \\
&= P\big(-W(1) > -x + o(1) \big) + o(1) \\
&\to \ \phi(x).
\end{aligned}
$$

(2.6.35)

$\square$

Now, we are in the position to prove of the second main theorem of this section.

Proof of Theorem 2.6.2. Lemma 2.6.15 gives us the required convergence towards the standard normal distribution, however it remains to convert $P(\tau \le H)$ into the expression given in Theorem 2.6.2. Since $w \overset{\text{a.s.}}{\simeq} h_2(k^*, \tilde{m}) \overset{\text{a.s.}}{\longrightarrow} \infty$, we have for sufficiently large m

$$P\big(\tau_{m,+}^{open} - k^* \le [w]\big) \ = \ P\big(\max\{\tau_{m,+}^{open} - k^*, 1\} \le [w] \big),$$

(2.6.36)

which by Lemma 2.6.6 and Lemma 2.6.15 implies that for sufficiently large m

$$P\big(\tau_{m,+}^{open} - k^* \le [w]\big) \ = \ P\big(r\big(\max\{\tau_{m,+}^{open} - k^*, 1\} \big) \le r([w]) \big).$$

An application of the mean value theorem allows us to replace $r([w])$ by $r(w) + \mathcal{O}(1)$ and the early-change assumption (2.6.3) allows us to replace $r(w(k^*))$ by $r(w(\tilde{m}))$, hence we obtain a standardization of the delay time independent of k^*. The latter is carried out in detail: By (2.6.27) we have

$$\frac{\sqrt{G(k^*)}}{g(k^*)} - \frac{\sqrt{G(\tilde{m})}}{g(\tilde{m})} \overset{\text{a.s.}}{=} o\big(\sqrt{m}\big).$$

Similar arguments as in the proof of (2.6.31) yield $h_2(k^*, \tilde{m})/h_2(\tilde{m}, \tilde{m}) \overset{\text{a.s.}}{=} 1 + o(1)$. Combining this with the mean value theorem, $h_2(\tilde{m}, \tilde{m}) \overset{\text{a.s.}}{\simeq} \sqrt{m}$ and the early-change assumption yields

$$\frac{h_2(k^*, \tilde{m})}{g(k^*)} - \frac{h_2(\tilde{m}, \tilde{m})}{g(\tilde{m})}$$
$$= \left(1 - \frac{h_2(\tilde{m}, \tilde{m})}{h_2(k^*, \tilde{m})}\right)\frac{h_2(k^*, \tilde{m})}{g(k^*)} + h_2(\tilde{m}, \tilde{m})\left(\frac{1}{g(k^*)} - \frac{1}{g(\tilde{m})}\right)$$
$$\overset{a.s.}{=} o(\sqrt{m}) + \mathcal{O}\big((k^* - \tilde{m})m^{1/2-\lambda-1}\big)$$
$$= o(\sqrt{m}) + \mathcal{O}\big(m^{\eta-\lambda-1/2}\big)$$
$$= o(\sqrt{m}).$$

The latter two assertions give us $r(w(k^*)) = r(w(\tilde{m})) + o(\sqrt{m})$. Plugging this in (2.6.36), solving the inequality for x and replacing δ by the consistent estimate $\hat{\delta}$ completes the proof for the stopping time $\tau_{m,+}^{open}$. The corresponding result for $\tau_{m,-}^{open}$ follows by similar arguments. $\qquad\square$

Proofs of the corollaries

Proof of Corollary 2.6.4. First of all, note that

$$\frac{\sqrt{G(\tilde{m})}\,\phi^{-1}(1-\tilde{\alpha}) + c\,h_1(\tilde{m}, T)}{|\delta|\,g(\tilde{m})/\sigma} \overset{a.s.}{\simeq} h_1(\tilde{m}, T) \overset{a.s.}{\longrightarrow} \infty,$$

hence q is a.s. well-defined for sufficiently large m and by Lemma 2.6.6 $q \overset{a.s.}{\longrightarrow} \infty$. So, for sufficiently large m we have

$$P\big(\tau_{m,+}^{closed} - k^* \le q\big) = P\big(\max\{\tau_{m,+}^{closed} - k^*, 1\} \le q\big).$$

Further, by the proof of Lemma 2.6.9 we have

$$P\big(\tau_{m,+}^{closed} \le k^*\big) \le P\Big(\max_{k=\tilde{m},\dots,k^*} |R_k|/h_1(k, T) > c\Big) \to 0.$$

Combining the assertions above yields

$$
\begin{aligned}
P\big(\tau_{m,+}^{closed} &- q \le k^* < \tau_{m,+}^{closed}\big) \\
&= P\big(k^* < \tau_{m,+}^{closed} \le k^* + q\big) \\
&= P\big(\tau_{m,+}^{closed} - k^* \le q\big) - P\big(\tau_{m,+}^{closed} \le k^*\big) \\
&= P\big(\max\{\tau_{m,+}^{closed} - k^*, 1\} \le q\big) + o(1).
\end{aligned}
$$

Plugging in the definition of q and applying Theorem 2.6.1 yields the first assertion. The second one follows in the same manner. On proving the third assertion one can copy the arguments of (1.5.17): If $\delta > 0$, we consider

$$
\begin{aligned}
P\big(\tau_m^{closed} &- q \le k^* < \tau_m^{closed}\big) \\
&\ge P\big(k^* < \tau_{m,+}^{closed} \le k^* + q\big) - P\big(\tau_{m,-}^{closed} \le k^*\big),
\end{aligned}
$$

where similar as above we obtain

$$P\big(k^* < \tau_{m,+}^{closed} \le k^* + q\big) \to 1 - \tilde{\alpha},$$

and

$$P\big(\tau_{m,-}^{closed} \le k^*\big) \to 0.$$

Exchanging the roles of $\tau_{m,+}^{closed}$ and $\tau_{m,-}^{closed}$ yields the assertion in case of $\delta < 0$. $\square$

Proof of Corollary 2.6.5. By Lemma 2.6.6 and Lemma 2.6.8 we know that $q(\chi_i)$ is a.s. well defined and $q(\chi_i) \overset{a.s.}{\longrightarrow} \infty$, for $i = 1, 2$. Thus, for sufficiently large m we have

$$
\begin{aligned}
P\big(\tau_{m,+}^{open} &- k^* \le q(\chi_1)\big) \\
&= P\big(\max\{\tau_{m,+}^{open} - k^*, 1\} \le q(\chi_1)\big) \\
&= P\big(r\big(\max\{\tau_{m,+}^{open} - k^*, 1\}\big) \le r(q(\chi_1))\big)
\end{aligned}
$$

and due to $\chi_1 = \phi^{-1}(1 - \tilde{\alpha} + \phi(-c)) > -c$ we can apply Theorem 2.6.2 as follows:

$$P\big(\tau_{m,+}^{open} - k^* \leq q(\chi_1)\big) \to \phi(\chi_1) = 1 - \tilde{\alpha} + \phi(-c). \tag{2.6.37}$$

By Lemma 2.6.13 and Lemma 2.6.14 we have

$$P\big(\tau_{m,+}^{open} \leq k^*\big)$$
$$= P\Big(\max_{k=\tilde{m},\dots,k^*} \frac{R_k}{h_2(k,\tilde{m})} > 1 \Big)$$
$$= P\Big(\frac{R_{k^*}}{\sqrt{G(k^*)}} + o_P(1) > \frac{h_2(k^*,\tilde{m})}{\sqrt{G(k^*)}} \Big).$$

Along the lines of Theorem 2.4.2 we have $R_{k^*}/\sqrt{G(k^*)} = \int_0^{k^*} g(x)\,dW(x)/\sqrt{G(k^*)} + o_P(1)$ and $\int_0^{k^*} g(x)\,dW(x)/\sqrt{G(k^*)} \overset{D}{=} W(1)$. Further, by the early-change assumption (2.6.3) we have $h_2(k^*,\tilde{m})/\sqrt{G(k^*)} \overset{a.s.}{=} c + o(1)$, which yields

$$P\big(\tau_{m,+}^{open} - k^* \leq 0\big) \to P(W(1) \geq c) = \phi(-c). \tag{2.6.38}$$

Combining (2.6.37) and (2.6.38) completes the proof for the one-sided, positive stopping time $\tau_{m,+}^{open}$. The proof for the one-sided, negative stopping time $\tau_{m,-}^{open}$ follows by the same arguments. On proving the assertion for the two-sided stopping time τ_m^{open} we copy the arguments of Corollary 1.5.3. $\qquad\square$

2.6.2   Asymptotic normality of the delay times for unknown in-control parameters

In this section we carry over the results of the previous section to the case of unknown in-control parameters.

Theorem 2.6.16. *With the notation and assumptions of Section 2.1 and Section 2.2 as well as $T = [\vartheta \, m^\rho]$ where*

$$1 < \rho < \frac{1/2 + \Lambda(1 - 2\beta)}{\Lambda(1 - 2\beta)}$$

and $\vartheta \geq 1$, δ being constant and being estimated by $\hat{\delta} = \delta + o_P\big(m^{(1-\rho)\Lambda(1/2-\beta)}\big)$, σ being estimated by $\hat{\sigma}_k = \hat{\sigma}_{\tilde{m}} = \sigma + o_P\big(m^{(1-\rho)\Lambda(1/2-\beta)}\big)$ for all $k = \tilde{m}, \ldots, \tilde{n}$, the critical value c being positive, $\lambda < \min\{1/2 - \kappa, 1/4\}$ and

$$k^* = m/\mu_0 + \mathcal{O}\big(m^\eta\big) \tag{2.6.39}$$

for some $\eta < 1 - \Lambda(\rho - 1)(1/2 - \beta)$ it holds for all $x \in \mathbb{R}$ under the alternative $H_{1,+}^{closed}$ that

$$\lim_{m \to \infty} P\left(\frac{\hat{\delta} \, \tilde{\lambda}_m \, r\big(\max\{\hat{\tau}_{m,+}^{closed} - k^*, 1\} \big)/\hat{\sigma}_{\tilde{m}} - c \, \hat{h}_1(\tilde{m}, T)}{\sqrt{\tilde{G}(\tilde{m})}} \leq x \right) = \phi(x)$$

and under the alternative $H_{1,-}^{closed}$ that

$$\lim_{m \to \infty} P\left(\frac{|\hat{\delta}| \tilde{\lambda}_m \, r\big(\max\{\hat{\tau}_{m,-}^{closed} - k^*, 1\} \big)/\hat{\sigma}_{\tilde{m}} - c \, \hat{h}_1(\tilde{m}, T)}{\sqrt{\tilde{G}(\tilde{m})}} \leq x \right) = \phi(x),$$

where $r(x)$ is defined in (2.6.1) and $\tilde{\lambda}_m = \lambda/(1 - \lambda) \, \tilde{m}^{-\lambda}$.

Theorem 2.6.17. *With the notation and assumptions of Section 2.1 and Section 2.2 as well as δ being constant and being estimated by $\hat{\delta} \xrightarrow{P} \delta$, σ being estimated by $\hat{\sigma}_k = \hat{\sigma}_{\tilde{m}} \xrightarrow{P} \sigma$ for all $k = \tilde{m}, \ldots, \tilde{n}$, the constant c from the threshold function $\hat{h}_2$ being positive, $\lambda < \min\{1/2 - \kappa, 1/4\}$ and*

$$k^* = m/\mu_0 + \mathcal{O}\big(m^\eta\big) \tag{2.6.40}$$

for some $\eta < \Lambda$ it holds for all $x > -c$ under the alternative $H_{1,+}^{open}$ that

$$\lim_{m \to \infty} P\left(\frac{\hat{\delta} \, \tilde{\lambda}_m \, r\big(\max\{\hat{\tau}_{m,+}^{open} - k^*, 1\} \big)/\hat{\sigma}_{\tilde{m}} - c\sqrt{\tilde{G}(\tilde{m})}}{\sqrt{\tilde{G}(\tilde{m})}} \leq x \right) = \phi(x)$$

and under the alternative $H_{1,-}^{closed}$ that

$$\lim_{m \to \infty} P\left(\frac{|\hat{\delta}| \, \tilde{\lambda}_m \, r\big(\max\{\hat{\tau}_{m,-}^{open} - k^*, 1\} \big)/\hat{\sigma}_{\tilde{m}} - c\sqrt{\tilde{G}(\tilde{m})}}{\sqrt{\tilde{G}(\tilde{m})}} \leq x \right) = \phi(x),$$

where $r(x)$ is defined in (2.6.1) and $\tilde{\lambda}_m = \lambda/(1 - \lambda) \, \tilde{m}^{-\lambda}$.

The following Lemma gives us a suitable estimate for δ, which fulfills the assumptions of Theorem 2.6.16 or Theorem 2.6.17, respectively.

Lemma 2.6.18. *With the notation and assumptions of Section 2.1 and Section 2.2 let*

$$\hat{\delta}_{m,k} = \frac{\sum_{i=1}^{k} \mu_{\tilde{m}}(i)\left(\tilde{S}_i - \tilde{S}_{i-1} - \tilde{S}_{\tilde{m}}/\tilde{m}\right)}{\sum_{i=1}^{k} \mu_{\tilde{m}}^2(i)},$$

where

$$\mu_l(k) := 1 - (1 + (k-l)_+)^{-\gamma} = \begin{cases} 1 - (1 + k - l)^{-\gamma} & \text{if } k > l, \\ 0 & \text{if } k \le l. \end{cases}$$

If $k^* - m = o(m^\zeta)$, as $m \to \infty$, for some $\zeta > 1$ it holds that, as $m \to \infty$

$$\hat{\delta}_{m,\tilde{m}+m^\zeta} \overset{a.s.}{=} \delta + \mathcal{O}(m^{1-\zeta}) + \mathcal{O}(m^{-\zeta\gamma}) + \mathcal{O}(\sqrt{\log\log(m)}m^{-\zeta/2}).$$

On choosing

$$\zeta > \max\left\{1 + (\rho - 1)\Lambda(1/2 - \beta), \frac{(\rho - 1)\Lambda(1/2 - \beta)}{\gamma}, (\rho - 1)\Lambda(1 - 2\beta)\right\}$$

the assumptions on $\hat{\delta}$ of Theorem 2.6.16 are fulfilled and on choosing $\zeta > 1$ the assumptions of Theorem 2.6.17 are fulfilled.

The proof of Lemma 2.6.18 is given at the very end of this section. The following two implications of Theorem 2.6.16 and Theorem 2.6.17 give us asymptotic confidence intervals for the change point:

Corollary 2.6.19. *Let the assumptions of Theorem 2.6.16 hold true. On setting*

$$\chi = r^{-1}\left(\frac{\sqrt{\tilde{G}(\tilde{m})}\,\phi^{-1}(1 - \tilde{\alpha}) + c\,\hat{h}_1(\tilde{m}, T)}{|\hat{\delta}|\tilde{\lambda}_m/\hat{\sigma}_{\tilde{m}}}\right),$$

where $\tilde{\alpha} \in (0,1)$, we obtain for $\hat{\tau}_{m,+}^{closed}$, $\hat{\tau}_{m,-}^{closed}$ and $\hat{\tau}_m^{closed}$ as in Section 2.2.2, that, as $m \to \infty$,

$$P\left(\hat{\tau}_{m,+}^{closed} - \chi \le k^* < \hat{\tau}_{m,+}^{closed}\right) = 1 - \tilde{\alpha} + o(1),$$
$$P\left(\hat{\tau}_{m,-}^{closed} - \chi \le k^* < \hat{\tau}_{m,-}^{closed}\right) = 1 - \tilde{\alpha} + o(1),$$
$$P\left(\hat{\tau}_m^{closed} - \chi \le k^* < \hat{\tau}_m^{closed}\right) \ge 1 - \tilde{\alpha} + o(1),$$

hold true under $H_{1,+}^{closed}$, $H_{1,-}^{closed}$ or H_1^{closed}, respectively.

Corollary 2.6.20. *Let the assumptions of Theorem 2.6.17 hold true. On setting*

$$q(\chi_i) = r^{-1}\left(\frac{\sqrt{\tilde{G}(\tilde{m})}\,\phi^{-1}(\chi_i) + \tilde{G}(\tilde{m})}{|\hat{\delta}|\,g(\tilde{m})\tilde{\lambda}_m/\hat{\sigma}_m}\right),$$

where

$$\chi_i = \begin{cases} \phi^{-1}\big(1 - \tilde{\alpha} - \phi(-c)\big) & \text{if } i = 1, \\ \phi^{-1}\big(1 - \tilde{\alpha} - 2\phi(-c)\big) & \text{if } i = 2, \end{cases}$$

with

$$\tilde{\alpha} \in \begin{cases} \big(\phi(-c),\, 1\big) & \text{for } i = 1, \\ \big(2\,\phi(-c),\, 1\big) & \text{for } i = 2, \end{cases}$$

we obtain for $\hat{\tau}_{m,+}^{open}$ *and* $\hat{\tau}_{m,-}^{open}$ *as in Section 2.2.2 and* $\tilde{\tau}_m^{open} = \min\{\hat{\tau}_{m,+}^{open}, \hat{\tau}_{m,+}^{open}\}$ *that, as* $m \to \infty$,

$$P\big(\hat{\tau}_{m,+}^{open} - q(\chi_1) \leq T^* < \hat{\tau}_{m,+}^{open}\big) = 1 - \tilde{\alpha} + o(1),$$
$$P\big(\hat{\tau}_{m,-}^{open} - q(\chi_1) \leq T^* < \hat{\tau}_{m,+}^{open}\big) = 1 - \tilde{\alpha} + o(1),$$
$$P\big(\tilde{\tau}_m^{open} - q(\chi_2) \leq T^* < \tilde{\tau}_m^{open}\big) \geq 1 - \tilde{\alpha} + o(1),$$

hold true under $H_{1,+}^{open}$, $H_{1,-}^{open}$ *or* H_1^{open}, *respectively.*

Corollary 2.6.19 and Corollary 2.6.20 can be shown along the lines of the proofs of Corollary 2.6.4 and Corollary 2.6.5, hence the proofs are omitted.

2.6.2.1   Proofs

We start by introducing some notation which will be used in this section frequently. Let

$$\hat{H} := k^* + [\hat{w}(k^*)] \tag{2.6.41}$$

with

$$\hat{w}(k^*) := r^{-1}\left(\frac{\sqrt{\tilde{G}(k^*)}\,\sigma}{|\delta|\,\tilde{\lambda}_m} \left(\frac{\hat{h}(k^*)}{\sqrt{\tilde{G}(k^*)}} + x \right) \right), \tag{2.6.42}$$

where $\hat{h}(t)$ shall denote either $c\,\hat{h}_1(t, T)$ or $\hat{h}_2(t, \tilde{m})$, depending on the setting we are working with. If not stated otherwise, throughout this section $\hat{w}$ shall denote $\hat{w}(k^*)$.

Remark 2.6.21. *Along the lines of Remark 2.6.7 we see that in the closed-end setting* $\hat{w}$ *is a.s. well-defined for any* $x \in \mathbb{R}$, *whereas in the open-end setting* $\hat{w}$ *is well-defined for any* $x > -c$, *where* c *is the critical value of the respective stopping time.*

To further shorten the notation let

$$\hat{\Lambda}_k := \sum_{i=1}^{k} \frac{g(i)\big(\mu(i) - \frac{1}{k}\sum_{j=1}^{k}\mu(j)\big)}{\sigma}, \tag{2.6.43}$$

$$\hat{A}(a, b) := \left(\max_{k=[a],\dots,[b]} \frac{\text{sgn}(\delta)\,\hat{R}_k}{\hat{h}(k)} - \frac{\text{sgn}(\delta)\,\hat{\Lambda}_{\hat{H}}}{\hat{h}(\hat{H})} \right) \frac{\hat{h}(k^*)}{\sqrt{\tilde{G}(k^*)}},$$

where again $\hat{h}(t)$ shall denote either $c\,\hat{h}_1(t, T)$ or $\hat{h}_2(t, \tilde{m})$.

Proof of Theorem 2.6.16

Theorem 2.6.16 follows in a similar manner as Theorem 2.6.1, yet, for one thing, we have to make sure that the estimate $\hat{\mu}_{0,k}$ does not cause any problems and, for another thing, we have to check that the respective rates still hold true for $\tilde{G}$, $\hat{h}_1$, $\hat{H}$ and $\hat{w}$ instead of G, h, H and w. We start with a technical lemma:

Lemma 2.6.22. *Under the assumptions of Theorem 2.6.16 we have, as* $m \to \infty$,

$$\hat{w} \stackrel{a.s.}{\simeq} m^\lambda \left(\tilde{G}(k^*)\right)^\beta \left(\tilde{G}(T/\mu_0)\right)^{\frac{1}{2}-\beta} \to \infty, \tag{2.6.44}$$

$$\hat{H} \stackrel{a.s.}{\sim} k^* \sim m/\mu_0, \tag{2.6.45}$$

$$\tilde{G}(\hat{H})/\tilde{G}(k^*) \stackrel{a.s.}{=} 1 + \mathcal{O}(\hat{w}/m), \tag{2.6.46}$$

$$\hat{h}_1(\hat{H},T)/\hat{h}_1(k^*,T) \stackrel{a.s.}{=} 1 + \mathcal{O}(\hat{w}/m), \tag{2.6.47}$$

$$\left(\operatorname{sgn}(\delta)\,c - \frac{\hat{\Lambda}_{\hat{H}}}{\hat{h}_1(\hat{H},T)}\right)\frac{\hat{h}_1(k^*,T)}{\sqrt{\tilde{G}(k^*)}} \stackrel{a.s.}{\longrightarrow} -\operatorname{sgn}(\delta)\,x, \tag{2.6.48}$$

where $\hat{H}$, $\hat{w}$ *and* $\hat{\Lambda}_k$ *are defined in* (2.6.41), (2.6.42) *and* (2.6.43).

Proof of Lemma 2.6.22. Recall that

$$w = r^{-1}\left(\frac{\sqrt{\tilde{G}(k^*)}\,\sigma}{|\delta|\,\tilde{\lambda}_m}\left(\frac{c\,\hat{h}_1(k^*,T)}{\sqrt{\tilde{G}(k^*)}} + x\right)\right).$$

Arguing as in (2.4.18) yields

$$\frac{\hat{h}_1(k^*,T)}{\sqrt{\tilde{G}(k^*)}} \stackrel{a.s.}{\simeq} \left(\frac{\tilde{G}(T/\mu_0)}{\tilde{G}(k^*)}\right)^{1/2-\beta} \to \infty, \tag{2.6.49}$$

hence by the definitions of $\hat{w}$ and $\tilde{\lambda}_m$ and by Lemma 2.6.6 we get

$$\hat{w} \stackrel{a.s.}{\simeq} \frac{\hat{h}_1(k^*,T)}{\tilde{\lambda}_m} \stackrel{a.s.}{\simeq} m^\lambda \left(\tilde{G}(k^*)\right)^\beta \left(\tilde{G}(T/\mu_0)\right)^{1/2-\beta},$$

i.e. (2.6.44) holds true. Combining this with the early-change assumption (2.6.39) yields

$$(\hat{H} - k^*)/k^* = \hat{w}/k^* \stackrel{a.s.}{=} \mathcal{O}\!\left(m^{\lambda-1+\Lambda\beta+\rho\Lambda(1/2-\beta)}\right) = o(1),$$

where the last equality follows from our assumptions on ρ. Further, by Lemma 2.2.3 and the mean value theorem we have for some $\xi \in [k^*,\hat{H}]$ that

$$\left|\frac{\tilde{G}(\hat{H})}{\tilde{G}(k^*)} - 1\right| = \frac{\tilde{G}'(\xi)\,(\hat{H} - k^*)}{\tilde{G}(k^*)} \stackrel{a.s.}{=} \mathcal{O}\!\left(\hat{w}\,m^{-2\lambda}\,k^{*-\Lambda}\right) = \mathcal{O}(\hat{w}/k^*),$$

hence (2.6.46) holds true. By $\hat{H} \stackrel{a.s.}{\sim} k^*$ and the growth rate of Corollary 2.3.3 we can show that the estimates up to $\hat{\mu}_{0,\hat{H}}$ converge towards the in-control mean

μ_0, even though they contain (some) observations made after the change point:

$$\max_{k=k^*,\ldots,\hat{H}} |\hat{\mu}_{0,k} - \mu_0|$$

$$\leq \max_{k=k^*,\ldots,\hat{H}} \left| \frac{\tilde{S}_k - \sum_{i=1}^k \mu(i)}{k} - \mu_0 + \frac{\sum_{i=1}^k \mu(i)}{k} \right|$$

$$\stackrel{\text{a.s.}}{=} \mathcal{O}\left(\sqrt{\log\log(k^*)/k^*}\right) + \max_{k=k^*,\ldots,\hat{H}} \frac{|\delta \sum_{i=k^*+1}^k (1 - (1 + i - k^*)^{-\gamma})|}{k} \quad (2.6.50)$$

$$= \mathcal{O}\left(\sqrt{\log\log(k^*)/k^*}\right) + \mathcal{O}(\hat{w}/k^*)$$

$$\stackrel{\text{a.s.}}{=} o(1).$$

Combining this with (2.6.46) (glimpsing at (2.4.18)) yields (2.6.47). On showing (2.6.48) we decompose the $\hat{\Lambda}_{\hat{H}}$ into one part corresponding to the case of known in-control parameters and another part corresponding to the estimation of μ_0:

$$\hat{\Lambda}_{\hat{H}} = \sum_{i=1}^{\hat{H}} \frac{g(i)\,(\mu(i) - \mu_0)}{\sigma} + \sum_{i=1}^{\hat{H}} \frac{g(i)\big(\mu_0 - \sum_{j=1}^{\hat{H}} \mu(j)/\hat{H}\big)}{\sigma}. \quad (2.6.51)$$

For the first part we make use of the asymptotic we derived for known in-control parameters (see (2.6.15)):

$$\sum_{i=1}^{\hat{H}} \frac{g(i)\,(\mu(i) - \mu_0)}{\sigma} = \frac{\delta\,g(k^*)\,r(\hat{w})}{\sigma} + \mathcal{O}\big(\hat{w}^{-\gamma} + m^{-\lambda-1}\hat{w}^2 + 1\big). \quad (2.6.52)$$

We approximate the second term in a similar manner. On the one hand, we have

$$\sum_{i=1}^{\hat{H}} g(i) \stackrel{\text{a.s.}}{=} \hat{H} - \frac{\hat{H}^{1-\lambda}}{1-\lambda} + \mathcal{O}(\hat{H}^{-\lambda}) \quad (2.6.53)$$

and, on the other hand,

$$\mu_0 - \frac{\sum_{j=1}^{\hat{H}} \mu(j)}{\hat{H}}$$

$$= -\delta\left(1 - \frac{\sum_{j=1}^{\hat{H}}(1 + (j - k^*)_+)^{-\gamma}}{\hat{H}}\right)$$

$$= -\delta\left(1 - \frac{k^* + \sum_{j=1}^{\hat{H}-k^*}(1+j)^{-\gamma}}{\hat{H}}\right)$$

$$\stackrel{\text{a.s.}}{=} -\delta\left(1 - \frac{k^* + (\hat{H} - k^*)^{1-\gamma}/(1-\gamma)}{\hat{H}} + \mathcal{O}\big(\hat{w}^{-\gamma}/\hat{H}\big)\right) \quad (2.6.54)$$

$$= -\delta\left(\frac{[\hat{w}]}{\hat{H}} - \frac{[\hat{w}]^{1-\gamma}}{\hat{H}\,(1-\gamma)}\right) + \mathcal{O}\big(\hat{w}^{-\gamma}/\hat{H}\big)$$

$$= \frac{-\delta\,r(\hat{w})}{\hat{H}} + \mathcal{O}\big(\hat{w}^{-\gamma}/\hat{H} + 1/\hat{H}\big),$$

which combines to

$$
\begin{aligned}
\hat{\Lambda}_{\hat{H}} &\stackrel{\text{a.s.}}{=} \frac{\delta\, r(\hat{w})}{\sigma}\left(g(k^*) - 1 + \frac{\hat{H}^{-\lambda}}{1 - \lambda}\right) + \mathcal{O}\big(\hat{w}^{-\gamma} + \hat{w}^2\, m^{-\lambda-1} + 1\big) \\
&= \frac{\delta\, r(\hat{w})}{\sigma}\left(\frac{\hat{H}^{-\lambda}}{1 - \lambda} - (1 + k^*)^{-\lambda}\right) + \mathcal{O}\big(\hat{w}^{-\gamma} + \hat{w}^2\, m^{-\lambda-1} + 1\big) \\
&= \frac{\delta\, r(\hat{w})}{\sigma\, \tilde{m}^\lambda}\left(\left(\frac{\tilde{m}}{\hat{H}}\right)^\lambda \frac{1}{1 - \lambda} - \left(\frac{\tilde{m}}{1 + k^*}\right)^\lambda\right) \\
&\quad + \mathcal{O}\big(\hat{w}^{-\gamma} + \hat{w}^2\, m^{-\lambda-1} + 1\big).
\end{aligned}
$$

By applying the mean value theorem and the early-change assumption (2.6.39) we have

$$
\left|(\tilde{m}/\hat{H})^\lambda - 1\right| \stackrel{\text{a.s.}}{=} \mathcal{O}\big((\hat{H} - \tilde{m})/m\big) \stackrel{\text{a.s.}}{=} \mathcal{O}\big((\hat{w} + m^\eta)/m\big) \tag{2.6.55}
$$

and

$$
\left|1 - (\tilde{m}/(1 + k^*))^\lambda\right| \stackrel{\text{a.s.}}{=} \mathcal{O}\big(m^{\eta-1}\big), \tag{2.6.56}
$$

which yields

$$
\begin{aligned}
\hat{\Lambda}_{\hat{H}} &\stackrel{\text{a.s.}}{=} \frac{\lambda\,\delta\, r(\hat{w})}{\sigma\,(1-\lambda)\,\tilde{m}^\lambda} + \mathcal{O}\big(\hat{w}^{-\gamma} + \hat{w}^2\, m^{-\lambda-1} + \hat{w}\, m^{\eta-1-\lambda} + 1\big) \\
&\stackrel{\text{a.s.}}{=} \frac{\delta\, \tilde{\lambda}_m\, r(\hat{w})}{\sigma} + o\big(\sqrt{\tilde{G}(k^*)}\big),
\end{aligned} \tag{2.6.57}
$$

where for the last equality we used that

1. $\tilde{\lambda}_m = \lambda/(1-\lambda)\,\tilde{m}^{-\lambda}$,

2. $\hat{w}^{-\gamma}/\sqrt{\tilde{G}(k^*)} \stackrel{\text{a.s.}}{\longrightarrow} 0$,

3. $\hat{w}^2\, m^{-1-\lambda}/\sqrt{\tilde{G}(k^*)} \stackrel{\text{a.s.}}{=} \mathcal{O}(m^{(\rho-1)\Lambda(1-2\beta)-1/2}) = o(1)$,

4. $\hat{w}\, m^{\eta-1-\lambda}/\sqrt{\tilde{G}(k^*)} \stackrel{\text{a.s.}}{=} \mathcal{O}(m^{\eta-1+\Lambda(1/2-\beta)(1-\rho)}) = o(1)$,

5. $1/\sqrt{\tilde{G}(k^*)} \to 0$.

Now, plugging the approximation of $\hat{\Lambda}_{\hat{H}}$ into the left hand side of (2.6.48) yields

$$
\begin{aligned}
&\left(\text{sgn}(\delta)\, c - \frac{\hat{\Lambda}_{\hat{H}}}{\hat{h}_1(\hat{H}, T)}\right) \frac{\hat{h}_1(k^*, T)}{\sqrt{\tilde{G}(k^*)}} \\
&= \frac{\text{sgn}(\delta)\, c\, \hat{h}_1(k^*, T)}{\sqrt{\tilde{G}(k^*)}} - \frac{\hat{\Lambda}_{\hat{H}}}{\sqrt{\tilde{G}(k^*)}}\, \frac{\hat{h}_1(k^*, T)}{\hat{h}_1(\hat{H}, T)} \\
&\stackrel{\text{a.s.}}{=} \frac{\text{sgn}(\delta)\, c\, \hat{h}_1(k^*, T)}{\sqrt{\tilde{G}(k^*)}} - \left(\frac{\delta\, \tilde{\lambda}_m\, r(\hat{w})}{\sigma\, \sqrt{\tilde{G}(k^*)}} + o(1)\right)\big(1 + \mathcal{O}(\hat{w}/m)\big) \\
&= \frac{\text{sgn}(\delta)\, c\, \hat{h}_1(k^*, T) - \delta\, \tilde{\lambda}_m\, r(\hat{w})/\sigma}{\sqrt{\tilde{G}(k^*)}} + o(1).
\end{aligned}
$$

Plugging the definition of $\hat{w}$ in the equation above yields (2.6.48). $\square$

Lemma 2.6.23. *Under the assumptions of Theorem 2.6.16 we have*

$$\hat{A}(\tilde{m}, k^*) \xrightarrow{P} -\infty.$$

Proof of Lemma 2.6.23. Let $1 < \tilde{\rho} < \rho$. By Theorem 2.4.2 we have

$$\max_{k=\tilde{m},\ldots,k^*} \frac{|\hat{R}_k|}{\hat{h}_1(k,T)}$$

$$\leq \max_{k=\tilde{m},\ldots,k^*} \frac{|\hat{R}_k|}{\hat{h}_1(k,m^{\tilde{\rho}})} \max_{k=\tilde{m},\ldots,k^*} \frac{\hat{h}_1(k,m^{\tilde{\rho}})}{\hat{h}_1(k,T)}$$

$$= \mathcal{O}_P(1)\, o_P(1)$$

$$= o_P(1).$$

Combining this with

$$\mathrm{sgn}(\delta)\, \hat{\Lambda}_{\hat{H}}/(\hat{h}_1(\hat{H},T)) \xrightarrow{\text{a.s.}} c > 0$$

(see (2.6.48)) and

$$\hat{h}_1(k^*,T)/\sqrt{\widehat{G}(k^*)} \xrightarrow{\text{a.s.}} \infty$$

(see (2.6.49)) yields the assertion.   □

Lemma 2.6.24. *Under the assumptions of Theorem 2.6.1 we have, as* $m \to \infty$,

$$\max_{k=k^*+1,\ldots,\hat{H}} \frac{|\hat{R}_k - \hat{R}_{k^*} - \hat{\Lambda}_k|}{\hat{h}_1(k,T)} = o_P\big(\sqrt{\widehat{G}(k^*)}/\hat{h}_1(k^*,T)\big).$$

Proof of Lemma 2.6.24. We decompose the expression into one part which corresponds to the case of known in-control parameters and a second part which corresponds to the estimation of μ_0: For $k = k^* + 1, \ldots, \hat{H}$ we have

$$\hat{R}_k - \hat{R}_{k^*} - \hat{\Lambda}_k = \frac{\sigma}{\hat{\sigma}_{\tilde{m}}}\big(R_k - R_{k^*} - \Lambda_k\big)$$

$$- \frac{\sigma}{\hat{\sigma}_{\tilde{m}}} \sum_{i=1}^{k} \frac{g(i)}{\sigma} \frac{\tilde{S}_k - M(k)}{k} + \frac{\sigma}{\hat{\sigma}_{\tilde{m}}} \sum_{i=1}^{k^*} \frac{g(i)}{\sigma} \frac{\tilde{S}_{k^*} - M(k)}{k^*}$$

$$= \frac{\sigma}{\hat{\sigma}_{\tilde{m}}}\big(R_k - R_{k^*} - \Lambda_k\big) \hspace{4cm} (2.6.58)$$

$$- \frac{\sigma}{\hat{\sigma}_{\tilde{m}}} \sum_{i=1}^{k} \frac{g(i)}{\sigma} \frac{\tilde{S}_k - M(k) - (\tilde{S}_{k^*} - M(k))k/k^*}{k}$$

$$- \frac{\sigma}{\hat{\sigma}_{\tilde{m}}} \sum_{i=k^*+1}^{k} \frac{g(i)}{\sigma} \frac{\tilde{S}_{k^*} - M(k^*)}{k^*},$$

where R_k and Λ_k are the counterparts of $\hat{R}_k$ and $\hat{\Lambda}_k$ in case of known in-control parameters and $M(k) = \sum_{i=1}^{k} \mu(i)$ denotes the mean of S_k (see Section 2.1). A careful examination of the calculations in the proof of Lemma 2.6.10 with $\varsigma = k^{*-q}$ for some $q \in (2\lambda, 1/2)$ yields that from

1. $k^{*-\lambda-1/2}\hat{w}/\sqrt{\tilde{G}(k^*)} \simeq m^{-1/2+(\rho-1)\Lambda(1/2-\beta)} \to 0$,

2. $k^{*\kappa}/\sqrt{\tilde{G}(k^*)} \simeq k^{*\kappa-\Lambda/2} \to 0$,

3. $\sqrt{k^*}\,\varsigma_m/\tilde{G}(k^*) \simeq m^{(2\lambda-q)/2} \to 0$,

4. $\sqrt{\hat{w}/\tilde{G}(k^*)} \simeq m^{((\rho-1)\Lambda(1/2-\beta)-1+3\lambda)/2} \to 0$

we obtain

$$\max_{k=k^*+1,\dots,\hat{H}} \frac{|R_k - R_{k^*} - \Lambda_k|}{\hat{h}_1(k,T)} = o_P\big(\sqrt{\tilde{G}(k^*)}/\hat{h}_1(k^*,T)\big)$$

(see (2.6.18)). Hence, we focus on the latter two summands of the right hand side of (2.6.58). Concerning the first of the two summands we have

$$\sum_{i=1}^{k} \frac{g(i)}{\sigma} \frac{\tilde{S}_k - M(k) - \big(\tilde{S}_{k^*} - M(k)\big)k/k^*}{k}$$
$$\leq \frac{\max\{\mu_0,\mu_1\}\,\big|\tilde{S}_k - M(k) - k/k^*\big(\tilde{S}_{k^*} - M(k^*)\big)\big|}{\sigma}$$
$$\leq \frac{\max\{\mu_0,\mu_1\}\,\big|\big(\tilde{S}_k - M(k)\big)\big(1 - k/k^*\big)\big|}{\sigma}$$
$$+ \frac{\max\{\mu_0,\mu_1\}\,\big|\tilde{S}_k - M(k) - \tilde{S}_{k^*} + M(k^*)\big|k}{\sigma\,k^*}.$$

By Corollary 2.3.3 and Lemma 2.6.22 we have

$$\max_{k=k^*+1,\dots,\hat{H}} \frac{\big|\big(\tilde{S}_k - M(k)\big)\big(1 - k/k^*\big)\big|}{\hat{h}_1(k,T)}$$
$$= \mathcal{O}_P\big(\sqrt{\log\log(k^*)/k^*}\,\hat{w}/\hat{h}_1(k^*,T)\big) \qquad (2.6.59)$$
$$= o_P\big(\sqrt{\tilde{G}(k^*)}/\hat{h}_1(k^*,T)\big),$$

where the last equality holds since our assumptions on λ and ρ yield

$$\sqrt{\log\log(k^*)/k^*}\,\hat{w}/\sqrt{\tilde{G}(k^*)}$$
$$= \sqrt{\log\log(k^*)}\,m^{(\rho-1)\Lambda(1/2-\beta)+\lambda-1/2}$$
$$\to 0.$$

Further, by the proof of Lemma 2.6.10 (see (2.6.17) onwards) and the calculations above we have

$$\max_{k=k^*+1,\dots,\hat{H}} \frac{\big|\tilde{S}_k - M(k) - (\tilde{S}_{k^*} - M(k^*))\big|}{\hat{h}_1(k,T)} = o_P\big(\sqrt{\tilde{G}(k^*)}/\hat{h}_1(k^*,T)\big).$$

Finally, similar as in (2.6.59) we have

$$\sum_{i=k^*+1}^{k} \frac{g(i)}{\sigma} \frac{\tilde{S}_{k^*} - M(k^*)}{k^*} = o_P\big(\sqrt{\tilde{G}(k^*)}/\hat{h}_1(k^*,T)\big).$$

Combining the assertions above with the consistency of $\hat{\sigma}_{\tilde{m}}$ completes the proof. $\square$

Lemma 2.6.25. *Let* $\breve{H} = \breve{H}(\epsilon) = k^* + [(1-\epsilon)\hat{w}]$. *Under the assumptions of Theorem 2.6.16 it holds for any* $\epsilon > 0$ *that, as* $m \to \infty$,

$$P\big(\hat{A}(k^* + 1, \hat{H}) = \hat{A}(\breve{H}, \hat{H})\big) \to 1.$$

Proof of Lemma 2.6.25. The proof is given for $\delta > 0$, yet follows by the same arguments if $\delta < 0$. Along the lines of (2.6.19) and (2.6.20) we have

$$\begin{aligned}
&P\big(\hat{A}(k^* + 1, \hat{H}) = \hat{A}(\breve{H}, \hat{H})\big) \\
&= P\bigg(\max_{k=k^*+1,\ldots,\breve{H}-1} \frac{\hat{R}_k - \hat{R}_{k^*}}{\hat{h}_1(k,T)} \le \frac{\hat{R}_{\breve{H}} - \hat{R}_{k^*}}{\hat{h}_1(\breve{H},T)} + o_P(1)\bigg),
\end{aligned} \tag{2.6.60}$$

where to both terms in the latter expression we can apply Lemma 2.6.24: For the first term we have

$$\max_{k=k^*+1,\ldots,\breve{H}-1} \frac{\hat{R}_k - \hat{R}_{k^*}}{\hat{h}_1(k,T)} = \max_{k=k^*+1,\ldots,\breve{H}-1} \frac{\hat{\Lambda}_k}{\hat{h}_1(k,T)} + o_P(1).$$

By the fact that $\hat{\Lambda}_k$ is increasing (see (2.5.4)) we have

$$\frac{\hat{\Lambda}_{\breve{H}-1}}{\hat{h}_1(\breve{H}-1,T)} \le \max_{k=k^*+1,\ldots,\breve{H}-1} \frac{\hat{\Lambda}_k}{\hat{h}_1(k,T)} \le \max_{k=k^*+1,\ldots,\breve{H}-1} \frac{\hat{\Lambda}_{\breve{H}-1}}{\hat{h}_1(k,T)}, \tag{2.6.61}$$

which, in combination with (2.6.47) and (2.6.48), yields

$$\max_{k=k^*+1,\ldots,\breve{H}-1} \frac{\hat{\Lambda}_k}{\hat{h}_1(k,T)} \stackrel{a.s.}{=} \frac{\hat{\Lambda}_{\breve{H}-1}}{\hat{h}_1(\hat{H},T)} + \mathcal{O}\big(\hat{w}/k^*\big).$$

Thus, by the fact that $\hat{w}/k^* \stackrel{a.s.}{\longrightarrow} 0$ we have

$$\max_{k=k^*+1,\ldots,\breve{H}-1} \frac{\hat{R}_k - \hat{R}_{k^*}}{\hat{h}_1(k,T)} = \frac{\hat{\Lambda}_{\breve{H}-1}}{\hat{h}_1(\hat{H},T)} + o_P(1)$$

and, similarly,

$$\frac{\hat{R}_{\breve{H}} - \hat{R}_{k^*}}{\hat{h}_1(k,T)} = \frac{\hat{\Lambda}_{\breve{H}}}{\hat{h}_1(\hat{H},T)} + o_P(1).$$

Plugging the two assertions above into in (2.6.60) yields

$$\begin{aligned}
&P\big(\hat{A}(k^* + 1, \hat{H}) = \hat{A}(\breve{H}, \hat{H})\big) \\
&= P\big(0 \le (\hat{\Lambda}_{\breve{H}} - \hat{\Lambda}_{\breve{H}-1})/\hat{h}_1(\hat{H},T) + o_P(1)\big),
\end{aligned} \tag{2.6.62}$$

hence we approximate the difference between the two deterministic terms: Making use of (2.6.52), (2.6.53) and (2.6.54) gives us

$$\begin{aligned}
&\hat{\Lambda}_{\breve{H}} - \hat{\Lambda}_{\breve{H}-1} \\
&= \sum_{i=\breve{H}}^{\hat{H}} \frac{g(i)\,(\mu(i) - \mu_0)}{\sigma} - \sum_{i=1}^{\hat{H}} \frac{g(i)}{\sigma} \sum_{j=1}^{\hat{H}} \frac{\mu(j) - \mu_0}{\hat{H}} \\
&\quad + \sum_{i=1}^{\breve{H}-1} \frac{g(i)}{\sigma} \sum_{j=1}^{\breve{H}-1} \frac{\mu(j) - \mu_0}{\breve{H} - 1}
\end{aligned}$$

$$\overset{\text{a.s.}}{=} \frac{\delta\, g(k^*) \left(r(\hat{w}) - r((1-\epsilon)\hat{w}) \right)}{\sigma} + \mathcal{O}\big(\hat{w}^{-\gamma} + m^{-\lambda-1}\hat{w}^2 + 1\big)$$

$$- \left(1 - \frac{\hat{H}^{-\lambda}}{1-\lambda} + \mathcal{O}\big(\hat{H}^{-\lambda-1}\big) \right) \left(\frac{\delta\, r(\hat{w})}{\sigma} + \mathcal{O}\big(\hat{w}^{-\gamma} + 1\big) \right)$$

$$+ \left(1 - \frac{(\check{H}-1)^{-\lambda}}{1-\lambda} + \mathcal{O}\big(\hat{H}^{-\lambda-1}\big) \right)$$

$$\times \left(\frac{\delta\, r((1-\epsilon)\hat{w})}{\sigma} + \mathcal{O}\big(\hat{w}^{-\gamma} + 1\big) \right)$$

$$= \frac{\delta\left(r(\hat{w}) - r((1-\epsilon)\hat{w}) \right)}{\sigma} \left(g(k^*) - 1 + \frac{\hat{H}^{-\lambda}}{1-\lambda} \right)$$

$$+ \frac{\hat{H}^{-\lambda} - (\check{H}-1)^{-\lambda}}{1-\lambda} \frac{\delta\, r((1-\epsilon)\hat{w})}{\sigma}$$

$$+ \mathcal{O}\big(\hat{w}^{-\gamma} + m^{-\lambda-1}\hat{w}^2 + 1\big)$$

$$= \frac{\delta\left(r(\hat{w}) - r((1-\epsilon)\hat{w}) \right)}{\tilde{m}^\lambda\, \sigma} \left(\left(\frac{\tilde{m}}{\hat{\check{H}}} \right)^\lambda \frac{1}{1-\lambda} - \left(\frac{\tilde{m}}{1+k^*} \right)^\lambda \right)$$

$$+ \mathcal{O}\big(\hat{w}^{-\gamma} + m^{-\lambda-1}\hat{w}^2\big)$$

$$\overset{\text{a.s.}}{=} \frac{\delta\, \lambda \left(r(\hat{w}) - r((1-\epsilon)\hat{w}) \right)}{\tilde{m}^\lambda\, \sigma\, (1-\lambda)}$$

$$+ \mathcal{O}\big(\hat{w}^{-\gamma} + m^{-\lambda-1}\hat{w}^2 + \hat{w}(\hat{w} + m^\eta)m^{-\lambda-1} + 1\big),$$

where the last equality is shown in (2.6.55) and (2.6.56). Further, by the definition of r we have

$$\frac{r(\hat{w}) - r((1-\epsilon)\hat{w})}{\epsilon\, \hat{w}} \overset{\text{a.s.}}{=} \frac{1}{\epsilon} - \frac{1-\epsilon}{\epsilon} + \mathcal{O}\big(\hat{w}^{-\gamma}\big) = 1 + \mathcal{O}\big(\hat{w}^{-\gamma}\big),$$

hence

$$\frac{\hat{\Lambda}_{\hat{H}} - \hat{\Lambda}_{\check{H}-1}}{\hat{h}_1(\hat{H}, T)} \overset{\text{a.s.}}{=} \frac{\delta\, \lambda\, \epsilon\, \hat{w}}{\sigma\, (1-\lambda)\, \tilde{m}^{-\lambda}\, \hat{h}_1(\hat{H}, T)} + o(1) \overset{\text{a.s.}}{\simeq} \frac{\delta\, \epsilon\, \lambda}{1-\lambda} + o(1).$$

Plugging the relation above into (2.6.62) yields the assertion. □

Lemma 2.6.26. *Under the assumptions of Theorem 2.6.16 we have under $H_{1,+}^{closed}$ respectivly $H_{1,-}^{closed}$ that*

$$\lim_{m\to\infty} P\big(\hat{\tau}_{m,+}^{closed} \leq \hat{H}\big) = \phi(x),$$

$$\lim_{m\to\infty} P\big(\hat{\tau}_{m,-}^{closed} \leq \hat{H}\big) = \phi(x).$$

Proof of Lemma 2.6.26. On combining Lemma 2.6.22, Lemma 2.6.23 and Lemma 2.6.25 we have

$$P\big(\hat{\tau}_{m,+}^{closed} \leq \hat{H}\big) = P\big(\hat{A}(\check{H}, \hat{H}) \geq -x + o(1)\big) + o(1).$$

Along the lines of the proof of Lemma 2.6.12, taking Lemma 2.6.24 and the fact that $\hat{\Lambda}_k$ is non-decreasing (see (2.5.4)) into account we see that

$$\hat{A}(\check{H}, \hat{H}) = \hat{R}_{k^*}/\sqrt{\tilde{G}(k^*)} + o_P(1).$$

By the arguments of the proof of Theorem 2.4.5 we have

$$\hat{R}_{k^*}/\sqrt{\tilde{G}(k^*)} = U(k^*)/\sqrt{\tilde{G}(k^*)} + o_P(1),$$

where U is defined in (2.4.21), and $U(k^*)/\sqrt{\tilde{G}(k^*)} \stackrel{D}{=} W(1)$. Hence, the assertion for the one-sided, positive stopping time $\hat{\tau}_{m,+}^{closed}$ holds true. The corresponding result for the one-sided, negative stopping time $\hat{\tau}_{m,-}^{closed}$ follows similarly (cf. also (2.6.24)). $\qquad\square$

Finally, we give the proof of the first main theorem of Section 2.6.2.

Proof of Theorem 2.6.16. By Lemma 2.6.26 and the arguments of the proof of Theorem 2.6.1 we have

$$P\Big(r\big(\max\{\hat{\tau}_{m,+}^{closed} - k^*, 1\}\big) \leq r([\hat{w}(k^*)])\Big) \to \phi(x). \tag{2.6.63}$$

In order to obtain a standardization of the delay time which is independent of k^* we aim to show that

$$r([\hat{w}(k^*)]) = r(\hat{w}(\tilde{m})) + o_P(\sqrt{\tilde{G}(k^*)}/\tilde{\lambda}_m). \tag{2.6.64}$$

By an application of the mean value theorem we obtain $r([\hat{w}(k^*)]) \stackrel{\text{a.s.}}{=} r(\hat{w}(k^*)) + \mathcal{O}(1)$. Further, by the early-change assumption it holds that

$$\left| \frac{\sqrt{\tilde{G}(k^*)} - \sqrt{\tilde{G}(\tilde{m})}}{\tilde{\lambda}_m} \right| \frac{\tilde{\lambda}_m}{\sqrt{\tilde{G}(k^*)}} \stackrel{\text{a.s.}}{=} o(1). \tag{2.6.65}$$

The mean value theorem and our assumptions on η yield for $0 < \beta < 1/2$ that

$$\begin{aligned}
&\left| \frac{\hat{h}_1(k^*, T)}{\tilde{\lambda}_m} - \frac{\hat{h}_1(\tilde{m}, T)}{\tilde{\lambda}_m} \right| \frac{\tilde{\lambda}_m}{\sqrt{\tilde{G}(k^*)}} \\
&= \mathcal{O}\big(T^{\Lambda(1/2-\beta)} m^{-\Lambda/2}\big) \big((\tilde{G}(k^*))^\beta - (\tilde{G}(\tilde{m}))^\beta\big) \\
&= \mathcal{O}\big(T^{\Lambda(1/2-\beta)} m^{-\Lambda/2}\big) m^\eta (\tilde{G}(\tilde{m}))^{\beta-1} \tilde{G}'(\tilde{m}) \\
&\stackrel{\text{a.s.}}{=} \mathcal{O}\big(T^{\Lambda(1/2-\beta)} m^{-\Lambda/2+\eta+\Lambda(\beta-1)-2\lambda}\big) \\
&= o(1),
\end{aligned}$$

where the same rate of convergence holds true if $\beta = 0$. Plugging (2.6.64) into (2.6.63) yields

$$P\left(\frac{|\delta| \tilde{\lambda}_m r\big(\max\{\hat{\tau}_{m,+}^{closed} - k^*, 1\}\big)/\sigma - c\hat{h}_1(\tilde{m}, T)}{\sqrt{\tilde{G}(\tilde{m})}} \leq x \right) \to \phi(x).$$

Finally,

$$
\begin{aligned}
\frac{|\delta|\,\tilde{\lambda}_m\, r\big(\max\{\hat{\tau}_{m,+}^{closed} - k^*, 1\}\big)}{\sqrt{\tilde{G}(\tilde{m})}} \\
&= \mathcal{O}_P\big(\hat{h}_1(\tilde{m}, T)/\sqrt{\tilde{G}(\tilde{m})}\big) \\
&= \mathcal{O}_P\big(m^{(\rho-1)\Lambda(1/2-\beta)}\big)
\end{aligned}
$$

and the convergence rates of $\hat{\delta}$ and $\hat{\sigma}$ allow us to replace the unknown parameters by the respective estimates, which completes the proof of the first assertion of Theorem 2.6.16. The second assertion follows by the same arguments. $\qquad\square$

Proof of Theorem 2.6.17

Lemma 2.6.27. *Under the assumptions of Theorem 2.6.17 we have, as* $m \to \infty$,

$$\hat{w} \overset{a.s.}{\simeq} m^\lambda \hat{h}_2(k^*, \tilde{m}) \overset{a.s.}{\longrightarrow} \infty, \tag{2.6.66}$$

$$\hat{H} \overset{a.s.}{\sim} k^* \sim m/\mu_0, \tag{2.6.67}$$

$$\tilde{G}(\hat{H})/\tilde{G}(k^*) \overset{a.s.}{=} 1 + \mathcal{O}(\hat{w}/k^*), \tag{2.6.68}$$

$$\hat{h}_2(\hat{H}, \tilde{m})/\hat{h}_2(k^*, \tilde{m}) \overset{a.s.}{=} 1 + \mathcal{O}(\hat{w}/k^*), \tag{2.6.69}$$

$$\left(\mathrm{sgn}(\delta) - \frac{\hat{\Lambda}_{\hat{H}}}{\hat{h}_2(\hat{H}, \tilde{m})} \right) \frac{\hat{h}_2(\hat{H}, \tilde{m})}{\sqrt{\tilde{G}(k^*)}} \overset{a.s.}{\longrightarrow} -\mathrm{sgn}(\delta)\, x. \tag{2.6.70}$$

Proof of Lemma 2.6.27. The first assertion follows along the lines of (2.6.28), taking $\tilde{\lambda}_m \simeq m^\lambda$ into account. The second assertion can be seen as follows: By Remark 1.7.1, Lemma 2.2.3 and (2.6.66) we have

$$\frac{\hat{w}}{k^*} \overset{a.s.}{\simeq} \frac{m^\lambda \hat{h}_2(k^*, \tilde{m})}{k^*} \sim \frac{m^\lambda \sqrt{\tilde{G}(k^*)}}{\sqrt{k^*}} \sqrt{\frac{\ln(\tilde{G}(k^*)/\tilde{G}(\tilde{m})) + f(c)}{k^*}} \to 0.$$

(2.6.68) follows along the lines of (2.6.46) and (2.6.69) follows along the lines of (2.6.31). So, (2.6.70) remains to be shown. Similar as in (2.6.57) we have

$$\hat{\Lambda}_{\hat{H}} \overset{a.s.}{=} \frac{\delta\, \tilde{\lambda}_m\, r(\hat{w})}{\sigma} + \mathcal{O}_P\big(\hat{w}^{-\gamma} + \hat{w}^2\, m^{-\lambda-1} + \hat{w}\, m^{\eta-1-\lambda} + 1\big)$$

$$\overset{a.s.}{=} \frac{\delta\, \tilde{\lambda}_m\, r(\hat{w})}{\sigma} + o_P\big(\sqrt{\tilde{G}(k^*)}\big),$$

where for the last equality we used

1. $\hat{w}^{-\gamma}/\sqrt{\tilde{G}(k^*)} \overset{a.s.}{=} o(1)$,

2. $\hat{w}^2\, m^{-1-\lambda}/\sqrt{\tilde{G}(k^*)} \overset{a.s.}{=} \mathcal{O}\big(m^{-1/2}\big) \overset{a.s.}{=} o(1)$,

3. $\hat{w}\, m^{\eta-1-\lambda}/\sqrt{\tilde{G}(k^*)} \overset{a.s.}{=} \mathcal{O}\big(m^{\eta-1}\hat{h}_2(k^*, \tilde{m})/\sqrt{\tilde{G}(k^*)}\big) \overset{a.s.}{=} o(1)$,

4. $1/\sqrt{\tilde{G}(k^*)} = o(1)$.

The approximation of $\hat{\Lambda}_{\hat{H}}$ gives us

$$\left(\mathrm{sgn}(\delta) - \frac{\hat{\Lambda}_H}{\hat{h}_2(\hat{H}, \tilde{m})} \right) \frac{\hat{h}_2(\hat{H}, \tilde{m})}{\sqrt{\tilde{G}(k^*)}}$$

$$\overset{a.s.}{=} \frac{\mathrm{sgn}(\delta)\, \hat{h}_2(\hat{H}, \tilde{m})}{\sqrt{\tilde{G}(k^*)}} - \frac{\delta\, \tilde{\lambda}_m\, r(\hat{w})}{\sigma\, \sqrt{\tilde{G}(k^*)}} + o(1)$$

$$= \frac{\mathrm{sgn}(\delta)\, \hat{h}_2(\hat{H}, \tilde{m})}{\sqrt{\tilde{G}(k^*)}} - \frac{\delta}{|\delta|} \left(\frac{\hat{h}_2(k^*, \tilde{m})}{\sqrt{\tilde{G}(k^*)}} + x \right) + o(1)$$

$$= \frac{\mathrm{sgn}(\delta)\, \hat{h}_2(\hat{H}, \tilde{m})}{\sqrt{\tilde{G}(k^*)}} \left(1 - \frac{\hat{h}_2(k^*, \tilde{m})}{\hat{h}_2(\hat{H}, \tilde{m})} \right) - \mathrm{sgn}(\delta)\, x + o(1)$$

$$\overset{a.s.}{=} -\mathrm{sgn}(\delta)\, x + o_P(1),$$

where in the last equality we used (2.6.69). $\qquad\qquad\qquad\qquad$ $\square$

Lemma 2.6.28. *Under the assumptions of Theorem 2.6.17 we have*

$$\max_{k=\tilde{m},\dots,\hat{H}} \frac{|\hat{R}_k - \hat{R}_{k^*} - \hat{\Lambda}_k|}{\hat{h}_2(k,\tilde{m})} = o_P(1).$$

Proof of Lemma 2.6.28. On replacing $\hat{h}_1$ by $\hat{h}_2$ the assertion for $k > k^*$ follows along the lines of Lemma 2.6.24, i.e.

$$\max_{k=k^*+1,\dots,\hat{H}} \frac{|\hat{R}_k - \hat{R}_{k^*} - \hat{\Lambda}_k|}{\hat{h}_2(k,\tilde{m})} = o_P(1).$$

For $k \leq k^*$ we have $\hat{\Lambda}_k = 0$, thus in this case we consider

$$\hat{R}_k - \hat{R}_{k^*} = \frac{\sigma}{\hat{\sigma}_{\tilde{m}}} (R_k - R_{k^*})$$

$$- \frac{\sigma}{\hat{\sigma}_{\tilde{m}}} \left(\sum_{i=1}^{k} \frac{g(i)(\tilde{S}_k/k - \mu_0)}{\sigma} - \sum_{i=1}^{k^*} \frac{g(i)(\tilde{S}_i/k^* - \mu_0)}{\sigma} \right).$$

For one thing, we have along the lines of Lemma 2.6.14, taking $\hat{h}_2(\tilde{m},\tilde{m}) \overset{\text{a.s.}}{\simeq} m^{\Lambda/2}$ and our assumptions on λ and η into account, that

$$\max_{k=\tilde{m},\dots,k^*} \frac{|R_k - R_{k^*}|}{\hat{h}_2(k,\tilde{m})} = o_P(1).$$

For another thing, making use of Proposition 1.3.2, Theorem 2.3.1 and Corollary 2.3.3 we have uniformly for $k = \tilde{m},\dots,k^*$

$$\sum_{i=1}^{k} \frac{g(i)(\tilde{S}_k/k - \mu_0)}{\sigma} - \sum_{i=1}^{k^*} \frac{g(i)(\tilde{S}_i/k^* - \mu_0)}{\sigma}$$

$$= \sum_{i=1}^{k} \frac{g(i)(\tilde{S}_k/k - \tilde{S}_{k^*}/k^*)}{\sigma} + \sum_{i=k+1}^{k^*} \frac{g(i)(\tilde{S}_{k^*}/k^* - \mu_0)}{\sigma}$$

$$\overset{\text{a.s.}}{=} \sum_{i=1}^{k} \frac{g(i)}{\sigma} \left(\frac{\tilde{S}_k - \tilde{S}_{k^*}}{k} + \frac{\tilde{S}_{k^*}}{k^*} \left(\frac{k^*}{k} - 1 \right) \right)$$

$$+ \mathcal{O}\big(m^{\eta-1/2}\sqrt{\log\log(m)}\big)$$

$$\overset{\text{a.s.}}{=} \sum_{i=1}^{k} \frac{g(i)(W(k) - W(k^*))}{k} + \mathcal{O}(m^{\kappa})$$

$$+ \mathcal{O}\big(m^{\eta-1/2}\sqrt{\log\log(m)}\big)$$

$$\overset{\text{a.s.}}{=} \mathcal{O}\big(m^{\eta/2}\sqrt{\log(m)} + m^{\kappa} + m^{\eta-1/2}\sqrt{\log\log(m)}\big).$$

Hence by $\hat{h}_2(\tilde{m}, \tilde{m}) \overset{\text{a.s.}}{\simeq} m^{\Lambda/2}$ it holds that

$$\max_{k=\tilde{m},\dots,k^*} \frac{\sum_{i=1}^{k} g(i)(\tilde{S}_k/k - \mu_0) - \sum_{i=1}^{k^*}(\tilde{S}_i/k^* - \mu_0)}{\sigma \, \hat{h}_2(k, \tilde{m})}$$
$$\overset{\text{a.s.}}{=} \mathcal{O}\big(m^{(\eta-\Lambda)/2} + m^{\kappa-\Lambda/2} + m^{\eta-1/2-\Lambda/2}\sqrt{\log\log(m)}\big)$$
$$= o(1),$$

which yields in combination with the consistency of $\hat{\sigma}_m$ that

$$\max_{k=\tilde{m},\dots,k^*} \frac{|\hat{R}_{\hat{H}} - \hat{R}_{k^*} - \hat{\Lambda}_k|}{\hat{h}_2(k, \tilde{m})} \overset{\text{a.s.}}{=} o(1)$$

and thus completes the proof. $\square$

Lemma 2.6.29. *Under the assumptions of Theorem 2.6.17 we have under $H_{1,+}^{open}$ respectively $H_{1,-}^{open}$ that*

$$\lim_{m\to\infty} P\Big(\hat{\tau}_{m,+}^{open} \leq \hat{H}\Big) = \phi(x)$$
$$\lim_{m\to\infty} P\Big(\hat{\tau}_{m,-}^{open} \leq \hat{H}\Big) = \phi(x).$$

Proof of Lemma 2.6.29. By Lemma 2.6.28 we have

$$\max_{k=\tilde{m},\dots,\hat{H}} \frac{\hat{R}_k}{\hat{h}_2(k, \tilde{m})} - \frac{\hat{\Lambda}_{\hat{H}}}{\hat{h}_2(\hat{H}, \tilde{m})} \qquad (2.6.71)$$
$$= \max_{k=\tilde{m},\dots,\hat{H}} \frac{\hat{R}_{k^*} - \hat{\Lambda}_k}{\hat{h}_2(k, \tilde{m})} - \frac{\hat{\Lambda}_{\hat{H}}}{\hat{h}_2(\hat{H}, \tilde{m})} + o_P(1),$$

where in the following we aim to replace the right hand side by $\hat{R}_{k^*}/\hat{h}_2(\hat{H}, \tilde{m})$:

1. By (2.6.69) we have

$$\max_{k=\tilde{m},\dots,\hat{H}} \left| \frac{\hat{R}_{k^*}}{\hat{h}_2(k^*, \tilde{m})} - \frac{\hat{R}_{k^*}}{\hat{h}_2(\hat{H}, \tilde{m})} \right| = \mathcal{O}_P\big(\hat{w}/k^*\big) = o_P(1).$$

2. Since $\hat{\Lambda}_k = 0$ for $k \leq k^*$ but $\hat{\Lambda}_k > 0$ for $k > k^*$ it holds that

$$\max_{k=\tilde{m},\dots,\hat{H}} \frac{\hat{\Lambda}_k}{\hat{h}_2(k, \tilde{m})} - \frac{\hat{\Lambda}_k}{\hat{h}_2(\hat{H}, \tilde{m})}$$
$$= \max_{k=k^*+1,\dots,\hat{H}} \frac{\hat{\Lambda}_k}{\hat{h}_2(k, \tilde{m})} - \frac{\hat{\Lambda}_k}{\hat{h}_2(\hat{H}, \tilde{m})}$$
$$\overset{\text{a.s.}}{=} o(1).$$

and by (2.6.69) and (2.6.70) we obtain

$$\max_{k=k^*+1,\dots,\hat{H}} \left| \frac{\hat{\Lambda}_k}{\hat{h}_2(k, \tilde{m})} - \frac{\hat{\Lambda}_k}{\hat{h}_2(\hat{H}, \tilde{m})} \right|$$
$$= \max_{k=k^*+1,\dots,\hat{H}} \frac{|\hat{\Lambda}_k|}{\hat{h}_2(\hat{H}, \tilde{m})} \left(\frac{\hat{h}_2(\hat{H}, \tilde{m})}{\hat{h}_2(k, \tilde{m})} - 1 \right)$$
$$\overset{\text{a.s.}}{=} \mathcal{O}\big(\hat{w}/k^*\big).$$

3. Since by (2.5.4) $\hat{\Lambda}_k$ is non-decreasing in k it holds that $\max_{k=\tilde{m},\dots,\hat{H}}(\hat{\Lambda}_k - \hat{\Lambda}_{\hat{H}}) = 0$.

Combining the three relations as in (2.6.34) and plugging the result into (2.6.71) yields

$$\max_{k=\tilde{m},\dots,\hat{H}} \frac{\hat{R}_k}{\hat{h}_2(k,\tilde{m})} - \frac{\hat{\Lambda}_H}{\hat{h}_2(\hat{H},\tilde{m})} = \frac{\hat{R}_{k^*}}{\hat{h}_2(\hat{H},\tilde{m})} + o_P(1).$$

Since by the early change assumption

$$\frac{\hat{h}_2(\hat{H},\tilde{m})}{\sqrt{\tilde{G}(k^*)}} \overset{\text{a.s.}}{\simeq} f^{-1}(\ln(\tilde{G}(\hat{H})/\tilde{G}(\tilde{m})) + f(c)\ ) \overset{\text{a.s.}}{=} \mathcal{O}(1)$$

it holds that

$$\frac{\hat{h}_2(\hat{H},\tilde{m})}{\sqrt{\tilde{G}(k^*)}} \max_{k=\tilde{m},\dots,\hat{H}} \frac{\hat{R}_k}{\hat{h}_2(k,\tilde{m})} - \frac{\hat{\Lambda}_{\hat{H}}}{\hat{h}_2(\hat{H},\tilde{m})} \overset{\text{a.s.}}{=} \frac{\hat{R}_{k^*}}{\sqrt{\tilde{G}(k^*)}} + o(1).$$

Along the lines of the proof of Theorem 2.4.6 we make use of the invariance principle to replace our observations by a Wiener process. The Gaussian analogue of the test statistic can be approximated by $U(k^*)/\sqrt{\tilde{G}(k^*)}$, where $\{U(t)\,|\,t \geq 0\}$ is defined in (2.4.21). Finally, the fact that $U(k^*)/\sqrt{\tilde{G}(k^*)} \overset{D}{=} W(1)$ and (2.6.32) yield

$$P\big(\hat{\tau}^{open}_{m,+} \leq \hat{H}\big) \to P\big(W(1) > -x\big) = \phi(x).$$

The assertion for the one-sided, negative stopping time $\hat{\tau}^{open}_{m,-}$ follows analogously (see also (2.6.35)). □

Now, we are in the position to formulate the proof of the main theorem:

Proof of Theorem 2.6.17. The arguments of the proof of Theorem 2.6.2 carry over if we ensure that

$$r^{-1}(\hat{w}(k^*)) \overset{\text{a.s.}}{=} r^{-1}(\hat{w}(\tilde{m})) + o\big(\sqrt{\tilde{G}(k^*)}/\tilde{\lambda}_m\big).$$

By the early-change assumption we have, for one thing,

$$\left| \frac{\sqrt{\tilde{G}(k^*)} - \sqrt{\tilde{G}(\tilde{m})}}{\tilde{\lambda}_m} \right| \frac{\tilde{\lambda}_m}{\sqrt{\tilde{G}(k^*)}} \overset{\text{a.s.}}{=} o(1)$$

and, for another thing,

$$\left| \frac{\hat{h}_2(k^*,\tilde{m})}{\tilde{\lambda}_m} - \frac{\hat{h}_2(\tilde{m},\tilde{m})}{\tilde{\lambda}_m} \right| \frac{\tilde{\lambda}_m}{\sqrt{\tilde{G}(k^*)}}$$
$$= \left| f^{-1}(\ln(\tilde{G}(k^*)/\tilde{G}(\tilde{m})) + f(c)) - c\,\sqrt{\tilde{G}(\tilde{m})/\tilde{G}(k^*)} \right|$$
$$\overset{\text{a.s.}}{=} o(1).$$

Combining this with the consistency of $\hat{\sigma}$ and $\hat{\delta}$ completes the proof for the stopping time $\tau^{open}_{m,+}$. The corresponding result for $\tau^{open}_{m,-}$ follows by similar arguments. □

Proof of Lemma 2.6.18

Proof of Lemma 2.6.18. Note that

$$\sum_{i=1}^{\tilde{m}+m^\zeta} \mu_{\tilde{m}}(i) = \sum_{i=1}^{\tilde{m}^\zeta} \left(1 - (1+i)^{-\gamma}\right) = m^\zeta + \mathcal{O}\left(m^{\zeta(1-\gamma)}\right)$$

and similarly

$$\sum_{i=1}^{\tilde{m}+m^\zeta} \mu_{\tilde{m}}^2(i) = m^\zeta + \mathcal{O}\left(m^{\zeta(1-\gamma)}\right),$$

$$\sum_{i=1}^{\tilde{m}+m^\zeta} \mu_{\tilde{m}}(i)\,\mu_{k^*}(i) = m^\zeta - k^* + \mathcal{O}\left(m^{\zeta(1-\gamma)}\right).$$

Making use of Remark 1.3.3 and Corollary 2.3.3 we obtain

$$
\begin{aligned}
\hat{\delta}_{m,\tilde{m}+m^\zeta} &= \frac{\sum_{i=\tilde{m}+1}^{\tilde{m}+m^\zeta} \mu_{\tilde{m}}(i)\left(M(i) - M(i-1) - M(\tilde{m})/\tilde{m}\right)}{\sum_{i=1}^{k} \mu_{\tilde{m}}^2(i)} \\
&\quad + \frac{\sum_{i=\tilde{m}+1}^{\tilde{m}+m^\zeta} \mu_{\tilde{m}}(i)\left(\tilde{S}_i - M(i) - \tilde{S}_{i-1} + M(i-1)\right)}{\sum_{i=1}^{k} \mu_{\tilde{m}}^2(i)} \\
&\quad - \frac{\sum_{i=\tilde{m}+1}^{\tilde{m}+m^\zeta} \mu_{\tilde{m}}(i)\left(\tilde{S}_{\tilde{m}} - M(\tilde{m})/\tilde{m}\right)}{\sum_{i=1}^{k} \mu_{\tilde{m}}^2(i)} \\
&\overset{\text{a.s.}}{=} \frac{\sum_{i=k^*+1}^{\tilde{m}+m^\zeta} \mu_{\tilde{m}}(i)\,\delta\,\mu_{k^*}(i)}{\sum_{i=1}^{k} \mu_{\tilde{m}}^2(i)} + \mathcal{O}\left(\sqrt{m^\zeta \log\log(m)}/m^\zeta\right) \\
&= \delta\,\frac{m^\zeta - k^*}{m^\zeta} + \mathcal{O}\left(m^{-\zeta\gamma}\right) + \mathcal{O}\left(\sqrt{\log\log(m)}\,m^{-\zeta/2}\right) \\
&= \delta + \mathcal{O}\left(m^{1-\zeta}\right) + \mathcal{O}\left(m^{-\zeta\gamma}\right) + \mathcal{O}\left(\sqrt{\log\log(m)}\,m^{-\zeta/2}\right),
\end{aligned}
$$

where in the last line we used $k^* - m = o\left(m^\zeta\right)$. $\qquad\square$

2.7   Finite sample behavior

In this section we present a few simulation results in order to give an idea of the
finite sample behavior of the presented monitoring procedures. We consider an (unobservable) process

$$S_k = \sum_{i=1}^{k} \varepsilon_i + \mu(i),$$

where $\{\varepsilon_i \,|\, i \in \mathbb{N}\}$ are independently, Exp(1)-distributed random variables and

$$\mu(k) = \mu_0 + \delta \left(1 - (1 + (k - k^*)_+)^{-\gamma}\right)$$

determines the drift of the process, which (possibly) changes at k^*. We analyze
the case of a "negative" change, i.e. the case where the mean drops from a level
μ_0 towards a lower level $\mu_1 = \mu_0 - \delta$ (where $\delta > 0$ is constant). The simulations
are carried out for different lengths of the training period, i.e. the number of observations of $N(t)$ which are available at the beginning of the monitoring procedure.
Note that the corresponding number of available observations of the inverse process
$\tilde{S}_n$ is $\tilde{m} \approx m/\mu_0$. So, for the sake of comparability all simulations are carried out
with $\mu_0 = 5$ and $\sigma^2 = 1$.

Due to the strict assumptions on λ and γ in case of unknown parameters (e.g.
$T^{1/2-\lambda-\gamma}/\sqrt{\log\log(T)} \to \infty$, as $m \to \infty$, see Theorem 2.5.3), the choice of λ and
γ is rather limited. Note however, that a small γ corresponds to a smooth, slowly
changing mean, whereas a large γ corresponds to a steep, almost abruptly changing
mean. Hence, we are interested in small values of γ and for the sake of clarity we
focus on the case of $\lambda = \gamma = 0.2$.

In case of unknown in-control parameters we estimate σ^2 (non-sequentially) by

$$\hat{\sigma}_k^2 = \hat{\sigma}_{\tilde{m}}^2 = \frac{1}{\hat{m}(\tilde{m} - \hat{m} + 1)} \sum_{j=\hat{m}}^{\tilde{m}} \left(\tilde{S}_j - \tilde{S}_{j-\hat{m}} - \hat{m}\tilde{S}_{\tilde{m}}/\tilde{m}\right)^2,$$

for all $k \geq \tilde{m}$, where we set $\hat{m} = \tilde{m}^{0.25}$ (see Remark 2.2.2). Further, for Figure 2.3
and Figure 2.5 we estimated the parameter δ by

$$\hat{\delta}_{m,\tilde{m}+m^\zeta} = \frac{\sum_{i=1}^{\tilde{m}+m^\zeta} \mu_{\tilde{m}}(i) \left(\tilde{S}_i - \tilde{S}_{i-1} - \tilde{S}_{\tilde{m}}/\tilde{m}\right)}{\sum_{i=1}^{\tilde{m}+m^\zeta} \mu_{\tilde{m}}^2(i)}$$

where we used $\zeta = 1.05$ (see Lemma 2.6.18), while for Figure 2.2 and Figure 2.4
we used the true value of δ. Each result is based on 5000 repetitions.

2.7.1   Closed-end setting

For the simulation results in the closed-end setting we fix a truncation point of $T = [m^{1.3}]$, whereas we consider different values for the parameter β of the threshold function, more precisely we consider $\beta = 0$, $\beta = 0.25$ and $\beta = 0.49$. Table 2.1 and Table 2.2 show the relative frequency of a false alarm for known and unknown in-control parameters for an asymptotic level of the tests of $\alpha = 5\%$ and $\alpha = 10\%$. We see that the asymptotic levels are well attained, however for β close to $1/2$ the tests become very conservative.

Tables 2.3 and Table 2.4 show the relative frequency of a correct detection of a change, where the change point is located in the middle of the time horizon in the sense that $\mu_0\, k^* = T/2$ and where we content ourselves with the case of $\alpha = 5\%$. We see that the procedures have power 1 as the training period increases.

In Figure 2.2 and Figure 2.3 we compare histograms of the standardized delay times with the density of the standard normal distribution (again for known or unknown in-control parameters, respectively). Note that the convergence results of Section 2.6 merely hold true under the early-change assumptions, hence we consider $k^* = m/\mu_0 + m^{0.3}$. As one would expect, we obtain a better fit in case of known in-control parameters, while in case of unknown in-control parameters the histograms have a slight tendency to the left.

	$\alpha = 5\%$			$\alpha = 10\%$		
m	$\beta = 0$	$\beta = 0.25$	$\beta = 0.49$	$\beta = 0$	$\beta = 0.25$	$\beta = 0.49$
100	.0336	.0296	.0034	.0750	.0628	.0088
250	.0438	.0366	.0080	.0900	.0788	.0174
500	.0448	.0420	.0112	.0936	.0826	.0254
1000	.0466	.0474	.0138	.0924	.0854	.0336
2000	.0462	.0482	.0150	.0980	.0954	.0374
3000	.0486	.0476	.0200	.0990	.0934	.0402

Table 2.1: Relative frequency of a false alarm under the null hypothesis for known in-control parameters for $T = m^{1.3}$ and $\lambda = 0.2$

	$\alpha = 5\%$			$\alpha = 10\%$		
m	$\beta = 0$	$\beta = 0.25$	$\beta = 0.49$	$\beta = 0$	$\beta = 0.25$	$\beta = 0.49$
100	.0036	.0312	.0068	.0770	.0896	.0198
250	.0380	.0310	.0078	.0784	.0778	.0202
500	.0376	.0472	.0104	.0916	.0856	.0274
1000	.0458	.0408	.0106	.0918	.0868	.0234
2000	.0410	.0470	.0138	.0976	.0928	.0274
3000	.0474	.0464	.0164	.0976	.0972	.0290

Table 2.2: Relative frequency of a false alarm under the null hypothesis for unknown in-control parameters for $T = m^{1.3}$, $\lambda = 0.2$ and $\hat{m} = m^{0.25}$

	$\beta = 0$			$\beta = 0.25$			$\beta = 0.49$		
m	$\delta = 1$	$\delta = 2$	$\delta = 3$	$\delta = 1$	$\delta = 2$	$\delta = 3$	$\delta = 1$	$\delta = 2$	$\delta = 3$
100	.6980	.9950	1	.6596	.9918	1	.3718	.9696	1
250	.9986	1	1	.9966	1	1	.9826	1	1
500	1	1	1	1	1	1	1	1	1
1000	1	1	1	1	1	1	1	1	1
2000	1	1	1	1	1	1	1	1	1
3000	1	1	1	1	1	1	1	1	1

Table 2.3: Relative frequency of a correct detection of a change with known in-control parameters for $T = m^{1.3}$, $k^* = [T/(2\mu_0)]$, $\alpha = 5\%$ and $\lambda = \gamma = 0.2$

	$\beta = 0$			$\beta = 0.25$			$\beta = 0.49$		
m	$\delta = 1$	$\delta = 2$	$\delta = 3$	$\delta = 1$	$\delta = 2$	$\delta = 3$	$\delta = 1$	$\delta = 2$	$\delta = 3$
100	.4286	.9586	.9998	.3814	.9428	1	.1212	.7530	.9908
250	.9446	1	1	.9278	1	1	.6804	1	1
500	1	1	1	1	1	1	.9988	1	1
750	1	1	1	1	1	1	1	1	1
1000	1	1	1	1	1	1	1	1	1
2000	1	1	1	1	1	1	1	1	1
3000	1	1	1	1	1	1	1	1	1

Table 2.4: Relative frequency of a correct detection of a change with unknown in-control parameters for $T = m^{1.3}$, $k^* = [T/(2\mu_0)]$, $\alpha = 5\%$, $\lambda = \gamma = 0.2$ and $\hat{m} = m^{0.25}$

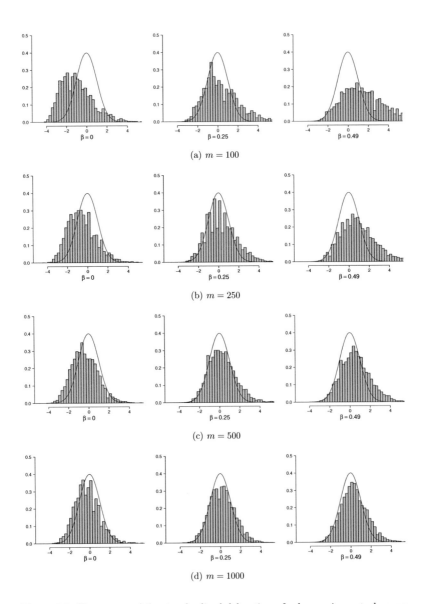

Figure 2.2: Histograms of the standardized delay times for known in-control parameters for $\lambda = \gamma = 0.2$, $\alpha = 5\%$, $\mu_1 = 3$, $\delta = 2$, $T = m^{1.3}$ and $k^* = m/\mu_0 + m^{0.3}$

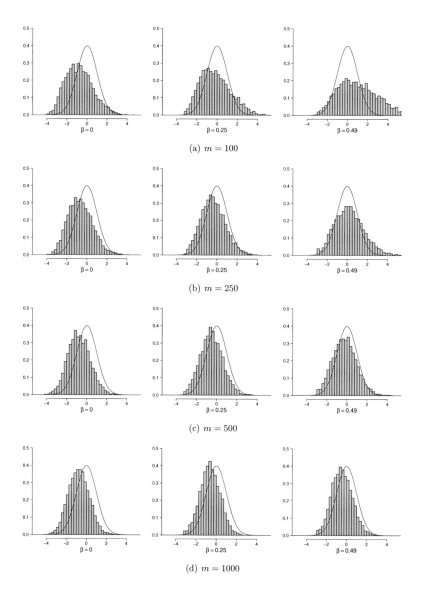

Figure 2.3: Histograms of the standardized delay times for unknown in-control parameters for $\lambda = \gamma = 0.2$, $\alpha = 5\%$, $\mu_1 = 3$, $\delta = 2$, $T = m^{1.3}$, $k^* = m/\mu_0 + m^{0.3}$, $\hat{m} = m^{0.25}$ and $\zeta = 1.05$

2.7.2 Open-end setting

For the open-end simulations we consider a time horizon of $T = 10\,m$ realizations of $N(t)$. Table 2.5 and Table 2.6 show the relative frequency of a false alarm for known and unknown in-control parameters, where the asymptotic level of the test is $\alpha = 5\%$ or $\alpha = 10\%$, respectively. The asymptotic levels are well attained, in fact, the procedures are quite conservative.

On illustrating the behavior of the monitoring procedures under the alternative, we consider the case of $k^* = m/\mu_0 + m^{0.3}$ which corresponds to the early-change setting under which the asymptotic behavior of the (standardized) delay times is known. Table 2.7 and Table 2.8 show the relative frequency of a correct detection of the change, where the histograms below (Figure 2.4 and Figure 2.5) compare the standardized delay times with the density function of the standard normal distribution. The results are quite satisfactory as m increases.

α \\ m	100	250	500	750	1000	2000	3000
5%	.0108	.0162	.0232	.0246	.0280	.0258	.0320
10%	.0344	.0560	.0572	.0602	.0662	.0556	.0682

Table 2.5: Relative frequency of a false alarm under the null hypothesis for known in-control parameters for $\lambda = 0.2$

α \\ m	100	250	500	750	1000	2000	3000
5%	.0262	.0256	.0272	.0338	.0289	.0322	.0344
10%	.0508	.0516	.0698	.0634	.0604	.0708	.0656

Table 2.6: Relative frequency of a false alarm under the null hypothesis for unknown in-control parameters for $\lambda = 0.2$ and $\hat{m} = m^{0.25}$

δ \\ m	100	250	500	750	1000	2000	3000
0.5	.8342	.9992	1	1	1	1	1
1	1	1	1	1	1	1	1
2	1	1	1	1	1	1	1
3	1	1	1	1	1	1	1

Table 2.7: Relative frequency of a correct detection of a change with known in-control parameters for $\alpha = 5\%$, $\lambda = \gamma = 0.2$ and $k^* = m/\mu_0 + m^{0.3}$

δ \ m	100	250	500	750	1000	2000	3000
0.5	.1624	.4130	0.8096	.9584	1	1	1
1	.6850	.9896	1	1	1	1	1
2	.9950	1	1	1	1	1	1
3	1	1	1	1	1	1	1

Table 2.8: Relative frequency of a correct detection of a change with unknown in-control parameters for $\alpha = 5\%$, $\lambda = \gamma = 0.2$, $\hat{m} = m^{0.25}$ and $k^* = m/\mu_0 + m^{0.3}$

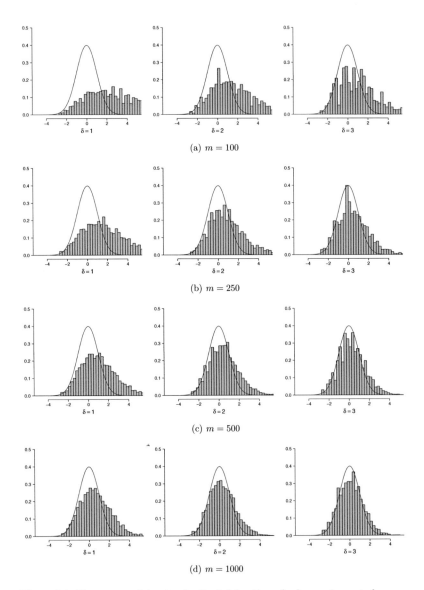

Figure 2.4: Histograms of the standardized delay times for known in-control parameters for $\lambda = \gamma = 0.2$, $\alpha = 5\%$ and $k^* = m/\mu_0 + m^{0.3}$

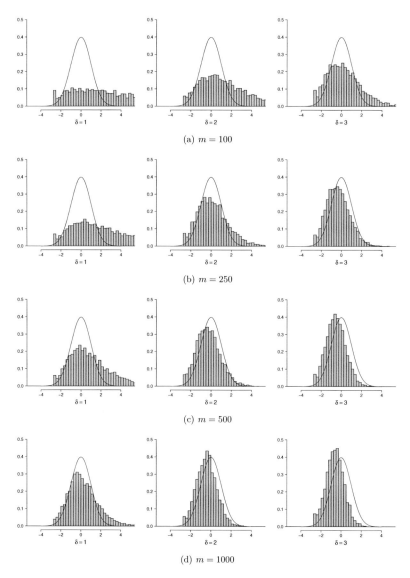

Figure 2.5: Histograms of the standardized delay times for unknown in-control parameters for $\lambda = \gamma = 0.2$, $\alpha = 5\%$, $k^* = m/\mu_0 + m^{0.3}$, $\hat{m} = m^{0.25}$ and $\zeta = 1.05$.

Literature

S. Asmussen. *Applied Probability and Queues*. Springer, New York, 2003.

A. Aue. *Sequential Change-Point Analysis Based on Invariance Principles*. PhD thesis, University of Cologne, 2003.

A. Aue and L. Horváth. Delay time in sequential detection of change. *Statistics and Probability Letters*, 67(3):221–231, 2004.

A. Aue and L. Horváth. Structural breaks in time series. *Journal of Time Series Analysis*, 34(1):1–16, 2013.

I. Berkes, E. Gombay, L. Horváth, and P. Kokoszka. Sequential change-point detection in GARCH(p,q) models. *Econometric Theory*, 20:1140–1167, 2004.

C.S.J. Chu, M. Stinchcombe, and H. White. Monitoring structural change. *Econometrika*, 64(5):1045–1065, 1996.

M. Csörgő and L. Horváth. *Weighted Approximations in Probability and Statistics*. Wiley, Chichester, 1993.

M. Csörgő and L. Horváth. *Limit Theorems in Change-Point Analysis*. Wiley, Chichester, 1997.

M. Csörgő and P. Révész. How big are the increments of a Wiener process? *The Annals of Probability*, 7(4):731–737, 1979.

M. Csörgő and P. Révész. *Strong Approximations in Probability and Statistics*. Academic Press, New York, 1981.

M. Csörgő, L. Horváth, and J. Steinebach. Invariance principles for renewal processes. *The Annals of Probability*, 15(4):1441–1460, 1987.

A. Gut. *Stopped Random Walks: Limit Theorems and Applications*. Springer, New York, 2009.

A. Gut. Renewal theory with a trend. *Statistics and Probability Letters*, 81(8):1292–1299, 2011.

A. Gut and J. Steinebach. Truncated sequential change-point detection based on renewal counting processes. *Scandinavian Journal of Statistics*, 29(4):693–719, 2002.

A. Gut and J. Steinebach. Truncated sequential change-point detection based on renewal counting processes II. *Journal of Statistical Planning and Inference*, 139(6):1921–1936, 2009.

L. Horváth. Strong approximations of renewal processes and their applications. *Acta Mathematica Hungarica*, 47(1-2):13–28, 1986.

L. Horváth. Detection of changes in linear sequences. *Annals of the Institute of Statistical Mathematics*, 49(2):271–283, 1997.

L. Horváth and J. Steinebach. Testing for changes in the mean or variance of a stochastic process under weak invariance. *Journal of Statistical Planning and Inference*, 91(2):365–376, 2000.

L. Horváth, M. Hušková, P. Kokoszka, and J. Steinebach. Monitoring changes in linear models. *Journal of Statistical Planning and Inference*, 126(1):225–251, 2004.

M. Hušková. Remarks on test procedures for gradual changes. In *"Asymptotic Methods in Probability and Statistics"* (B. Szyszkowicz, ed.), pages 577–583. Elsevier, Amsterdam, 1998a.

M. Hušková. Estimators in the location model with gradual changes. *Commentationes Mathematicae Universitatis Carolinae*, 39(1):147–157, 1998b.

M. Hušková. Gradual changes versus abrupt changes. *Journal of Statistical Planning and Inference*, 76(1-2):109–125, 1999.

M. Hušková and J. Steinebach. Limit theorems for a class of tests of gradual changes. *Journal of Statistical Planning and Inference*, 89(1):57–77, 2000.

D. Jarušková. Testing appearance of linear trend. *Journal of Statistical Planning and Inference*, 70(2):263–276, 1998.

I. Karatzas and S. Shreve. *Brownian Motion and Stochastic Calculus*. Springer, New York, 1991.

C. Kirch. *Resampling Methods for the Change Analysis of Dependent Data*. PhD thesis, University of Cologne, 2006.

C. Kirch and J. Steinebach. Permutation principles for the change analysis of stochastic processes under strong invariance. *Journal of Computational and Applied Mathematics*, 186(1):64–88, 2006.

J. Komlós, P. Major, and G. Tusnády. An approximation of partial sums of independent r.v.'s and the sample df. I. *Zeitschrift für Wahrscheinlichkeitstheorie und verwandte Gebiete*, 32(1-2):111–131, 1975.

J. Komlós, P. Major, and G. Tusnády. An approximation of partial sums of independent r.v.'s, and the sample df. II. *Zeitschrift für Wahrscheinlichkeitstheorie und verwandte Gebiete*, 34(1):33–58, 1976.

M. Kühn. *Sequential Change–Point Analysis Based on Weighted Moving Averages*. PhD thesis, University of Cologne, 2008.

S. Mihalache. Strong approximations and sequential change-point analysis for diffusion processes. *Statistics and Probability Letters*, 82(3):464–472, 2012.

E.S. Page. Continuous inspection schemes. *Biometrika*, 41(1-2):100–115, 1954.

E. Parzen. *Stochastic Processes*. Society for Industrial and Applied Mathematics, Philadelphia, 1999.

H. Robbins and D. Siegmund. Boundary crossing probabilities for the Wiener process and sample sums. *The Annals of Mathematical Statistics*, 41(5):1410–1429, 1970.

J. Steinebach. Some remarks on the testing of smooth changes in the linear drift of a stochastic process. *Theory of Probability and Mathematical Statistics*, 61:173–185, 2000.

J. Steinebach and H. Timmermann. Sequential testing of gradual changes in the drift of a stochastic process. *Journal of Statistical Planning and Inference*, 141(8): 2682–2699, 2011.

V. Strassen. An invariance principle for the law of the iterated logarithm. *Zeitschrift für Wahrscheinlichkeitstheorie und verwandte Gebiete*, 3(3):211–226, 1964.

H. Timmermann. *Monitoring-Verfahren für stochastische Prozesse mit graduellen Strukturbrüchen*. Diploma thesis, University of Cologne, 2010.